AF389488

Surface Engineering of C/N/O Functionalized Materials

Surface Engineering of C/N/O Functionalized Materials

Editor

Yanxiang Zhang

MDPI • Basel • Beijing • Wuhan • Barcelona • Belgrade • Manchester • Tokyo • Cluj • Tianjin

Editor
Yanxiang Zhang
School of Materials Science and
Engineering
Harbin Institute of Technology
Harbin
China

Editorial Office
MDPI
St. Alban-Anlage 66
4052 Basel, Switzerland

This is a reprint of articles from the Special Issue published online in the open access journal *Coatings* (ISSN 2079-6412) (available at: www.mdpi.com/journal/coatings/special_issues/Surface_Engineering_Materials).

For citation purposes, cite each article independently as indicated on the article page online and as indicated below:

LastName, A.A.; LastName, B.B.; LastName, C.C. Article Title. *Journal Name* **Year**, *Volume Number*, Page Range.

ISBN 978-3-0365-2705-5 (Hbk)
ISBN 978-3-0365-2704-8 (PDF)

© 2021 by the authors. Articles in this book are Open Access and distributed under the Creative Commons Attribution (CC BY) license, which allows users to download, copy and build upon published articles, as long as the author and publisher are properly credited, which ensures maximum dissemination and a wider impact of our publications.
The book as a whole is distributed by MDPI under the terms and conditions of the Creative Commons license CC BY-NC-ND.

Contents

About the Editor

Yanxiang Zhang

Yanxiang Zhang is a Professor in the School of Materials Science and Engineering, Harbin Institute of Technology, Harbin, China. He received his B. E. in Materials Science from Harbin Institute of Technology in 2008 and Ph.D. in Materials Science from University of Science & Technology of China in 2013. His research interests include the theoretical modeling, process characterization, and technical development of heat treatment and surface engineering for metallic materials and functional ceramic materials. Prof. Zhang is serving as the Deputy Secretary General of China Heat Treatment Association and is an invited reviewer for National Natural Science Foundation of China (NSFC). He is currently an Associate Editor for e-Prime (Elsevier) and an Editor for Coatings (MDPI). Under the support of more than 20 projects, including NSFC, Ministry of National Defense Project of China, National Key Research & Development Program of China, Prof. Zhang developed a series of theories and methods for the characterization of 3D microstructures, the in situ characterization of reaction–diffusion process, and the applications in case hardening of gearing and tooling parts and analysis of ceramic fuel cells. He has published more than 100 peer-reviewed scientific papers with more than 2000 citations and H-index 26.

Preface to "Surface Engineering of C/N/O Functionalized Materials"

Great scientific and technological progress on heat treatment and surface modification has been achieved in the respective material categories to date, but cross references across material categories are barely made, although common technical mechanisms existing therein. The aim of this Special Issue is to present technical synergies, such as characterization and testing methodology, surface reaction mechanism, diffusion mechanism, process–structure–property relationships, etc., between the surface engineering of metals, ceramics, and polymers by evaluating the reaction–diffusion of C/N/O species, and to accelerate scientific discovery in the area of heat treatment and surface engineering. For example, the carburizing and nitriding of metallic materials are vital to enhancing the fatigue life of key base components, such as tools and dies; highly efficient oxygen transport in ceramic oxides is important to accelerate the practical applications of ceramic fuel cells. It appears that metallic materials and ceramic fuel cell materials belong to structural and functional materials, respectively. However, the thermochemical treatment of metallic materials, such as carburization and nitridation, and oxygen transport in ceramic fuel cells have similarities. They have a similar temperature range (400–1000 °C); the element C/N/O penetrates into the material bulk; the elemental processes are the same, i.e., reaction first then diffusion. Therefore, the two categories of materials can be integrated as "C/N/O Functionalized Materials". This Special Issue discusses the latest development of surface engineering of C/N/O functionalized materials, including both experimental and theoretical studies on heat treatment and surface engineering of metals, ceramics, and polymers.

The fundamental understanding of the kinetic reaction–diffusion process in materials depends on robust and precise methods for measurement of the kinetic parameters, such as surface exchange coefficient (k) and bulk diffusion coefficient (D). A typical method is the so-called "Electrical Conductivity Relaxation (ECR) Measurement". In recent years, the ECR method has been widely used for the mixed-ionic-electronic-conducting materials, such as the electrode materials for solid oxide fuel cells (SOFCs). However, the robustness and accuracy of measurement remains a challenge, since the rate-limiting mechanism and the experimental imperfections are not reflected explicitly in the time-domain ECR data. Yan et al. [1] developed a new method, called "the distribution of characteristic times (DCT)". Using the DCT spectrum, the rate-limiting mechanism and the refection of experimental imperfections are visualized clearly, and the values of k and D can be determined. A strong robustness of the DCT method is verified using noise-containing ECR data. The DCT method is in principle applicable to other materials, such as metals and polymers, by the proper relaxation measurement of the reaction–diffusion process.

For metallic materials, nitriding and carburizing are the best known and most widely applied surface engineering processes (also known as thermochemical treatment) for case hardening. Through the design of processing parameters, the surface layer microstructures can be refined, and therefore the corrosion, wear and fatigue properties of metals can be improved sufficiently. Yan et al. [2] proposed a novel decarburizing–nitriding treatment of low-carbon M50NiL and high-carbon M50-bearing steels. They showed that pre-decarburization reduces the activation energy for nitrogen diffusion and enhances nitrogen diffusivity, and the pre-decarburization can refine the surface-layer microstructure via a spinodal decomposition during plasma nitriding. Liu et al. [3] proposed a composite process of cold rolling and low-temperature plasma nitriding to improve the low hardness and poor wear resistance of the TA2 alloy. They showed that the wear property and hardness of deformed alloy samples can be improved by nitriding due to microstructural refinement. You et al. [4-5] studied the low-temperature plasma nitriding of 3Cr13 steel accelerated by rare-earth block, and the acceleration of plasma nitriding at 550 °C with rare earth on the surface of 38CrMoAl steel. They showed that most of the surface microstructures of the nitrided layer were refined by the addition of La. The presence of La reduces the N content in the modified layer, which accelerates the diffusion of N atoms and thus accelerates the nitriding process, and so the corrosion resistance is improved. Liu et al. [6] studied the effects of transition metal oxides (Ti, Zr, Nb, and Ta) on the mechanical properties and interfaces of B4C ceramics fabricated via pressureless sintering. They showed that the Ta_2O_5-added sample exhibited better elastic modulus, flexural strength, Vickers hardness, and fracture toughness, and exhibits the best combined properties when the mass fraction of the second phase was around five percent.

For ceramic materials, this Special Issue is restricted to the ceramic oxides for SOFCs, with the functionality (energy conversion from fuel to electricity) given by efficient ionic conduction and high catalytic activity. Ma et al. [7] studied the electrochemical activity of Fe-based perovskite cathodes for SOFCs. A novel cobalt-free perovskite oxide, $BaFe_{1-x}Y_xO_{3-\delta}$, was evaluated as the oxygen reduction electrode. Through the analysis of the distribution of relaxation times, the oxygen adsorption–dissociation process is determined to be the rate-limiting step at the electrode interface. In addition, a single cell with x=0.10 exhibits a good long-term stability. Yuan et al. [8] studied the effect of NiO addition to $La_{0.99}Ca_{0.01}NbO_4$ proton-conducting ceramic oxides for SOFCs. The NiO were added by directly mixing or by doping. They showed that both strategies improve the sinterability and conductivity, but the effect of doping is more significant in enhancing both grain growth and conductivity, making it more desirable for practical applications. The optimal doping amount of NiO was shown as 1~2 wt.%. The origin of the enhanced performance, revealed by first-principle calculations, is the decrease in both oxygen formation energy and hydration energy.

For polymer materials, Li et al. [9] studied the mechanical properties for basalt fiber/epoxy resin composites modified with La. To improve the poor interfacial adhesion between basalt fibers and the resin matrix, the modification solution containing different concentrations of Lanthanum ions was synthesized to modify the basalt fiber surfaces. They showed that La in the rare earth modification solution could link active oxygen-containing functional groups to the fibers'surfaces, and thus improve the roughness and the activity of the fiber surfaces, therefore enhancing the bonding between the resin matrix and fibers. In a following study [10], the effect of inorganic reinforced materials ($AgNO_3$, $FeCl_3 \cdot 6H_2O$, nano-graphene) on the mechanical and piezoelectric properties of electrospun PVDF fiber membranes was studied. The results showed that all the three materials can effectively promote the formation of the β-phase and thus enhance the piezoelectric performance. The best mechanical and piezoelectric properties were achieved by the addition of 1.0 wt.% nano-graphene and 0.3 wt.% AgNO3. Ke et al. [11] studied the fabrication process and properties of electrospun and electrosprayed polyethylene glycol/polylactic acid (PEG/PLA) films. PEG was introduced to enhance the cooling performance, due to its lower glass transition and melting temperatures. They showed that, the PEG/PLA film with a PLA content of 35 wt.% has a reduced thermal conductivity of 0.2 Wm-1K-1 and largest elasticity modulus of 378.3±68.5 MPa and tensile strength of 10.5±1.1 MPa.

I would like to express my greatest gratitude to the authors, the reviewers, and the editorial staff members who contributed enthusiastically to this Special Issue.

References

[1] Yan, F.; Wang, Y.; Yang, Y.; Zhu, L.; Hu, H.; Tang, Z.; Zhang, Y.; Yan, M.; Xia, C.; Xu, Y. Distribution of Characteristic Times: A High-Resolution Spectrum Approach for Visualizing Chemical Relaxation and Resolving Kinetic Parameters of Ionic-Electronic Conducting Ceramic Oxides. Coatings 2020, 10(12), 1240.

[2] Yan, F.; Yao, J.; Chen, B.; Yang, Y.; Xu, Y.; Yan, M.; Zhang, Y. A Novel Decarburizing-Nitriding Treatment of Carburized/through-Hardened Bearing Steel towards Enhanced Nitriding Kinetics and Microstructure Refinement. Coatings 2021, 11(2), 112.

[3] Liu, G.; Sun, H.; Wang, E.; Sun, K.; Zhu, X.; Fu, Y. Effect of Deformation on the Microstructure of Cold-Rolled TA2 Alloy after Low-Temperature Nitriding. Coatings 2021, 11(8), 1011.

[4] You, Y.; Li, R.; Yan, M.; Yan, J.; Chen, H.; Wang, C.; Liu, D.; Hong, L.; Han, T. Low-Temperature Plasma Nitriding of 3Cr13 Steel Accelerated by Rare-Earth Block. Coatings 2021, 11(9), 1050.

[5] Liu, D.; You, Y.; Yan, M.; Chen, H.; Li, R.; Hong, L.; Han, T. Acceleration of Plasma Nitriding at 550 °C with Rare Earth on the Surface of 38CrMoAl Steel. Coatings 2021, 11(9), 1122.

[6] Liu, G.; Chen, S.; Zhao, Y.; Fu, Y.; Wang, Y. The Effects of Transition Metal Oxides (Me = Ti, Zr, Nb, and Ta) on the Mechanical Properties and Interfaces of B4C Ceramics Fabricated via Pressureless Sintering. Coatings 2020, 10(12), 1253.

[7] Ma, D.; Gao, J.; Xia, T.; Li, Q.; Sun, L.; Huo, L.; Zhao, H. Insights in to the Electrochemical Activity of Fe-Based Perovskite Cathodes toward Oxygen Reduction Reaction for Solid Oxide Fuel Cells. Coatings 2020, 10(12), 1260.

[8] Yuan, K.; Liu, X.; Bi, L. Exploring the Effect of NiO Addition to $La_{0.99}Ca_{0.01}NbO_4$ Proton-Conducting Ceramic Oxides. Coatings 2021, 11(5), 562.

[9] Li, C.; Wang, H.; Zhao, X.; Fu, Y.; He, X.; Song, Y. Investigation of Mechanical Properties for Basalt Fiber/Epoxy Resin Composites Modified with La. Coatings 2021, 11(6), 666.

[10] Li, C.; Wang, H.; Yan, X.; Chen, H.; Fu, Y.; Meng, Q. Enhancement Research on Piezoelectric Performance of Electrospun PVDF Fiber Membranes with Inorganic Reinforced Materials. Coatings 2021, 11, 1495.

[11] Ke, W.; Li, X.; Miao, M.; Liu, B.; Zhang, X.; Liu, T. Fabrication and Properties of Electrospun and Electrosprayed Polyethylene Glycol/Polylactic Acid (PEG/PLA) Films. Coatings 2021, 11(7), 790.

Yanxiang Zhang

Editor

 **coatings**

Article

Distribution of Characteristic Times: A High-Resolution Spectrum Approach for Visualizing Chemical Relaxation and Resolving Kinetic Parameters of Ionic-Electronic Conducting Ceramic Oxides

Fuyao Yan [1], **Yiheng Wang** [1], **Ying Yang** [1], **Lei Zhu** [1], **Hui Hu** [1], **Zhuofu Tang** [1], **Yanxiang Zhang** [1,*], **Mufu Yan** [1,*], **Changrong Xia** [2,*] **and Yueming Xu** [3]

[1] National Key Laboratory for Precision Hot Processing of Metals, MIIT Key Laboratory of Advanced Structure-Function Integrated Materials and Green Manufacturing Technology, School of Materials Science and Engineering, Harbin Institute of Technology, Harbin 150001, China; fyan@hit.edu.cn (F.Y.); 20S009050@stu.hit.edu.cn (Y.W.); 15645040283@163.com (Y.Y.); 19S009069@stu.hit.edu.cn (L.Z.); 20S109221@stu.hit.edu.cn (H.H.); tangzhuofu0118@163.com (Z.T.)

[2] CAS Key Laboratory of Materials for Energy Conversion, Department of Materials Science and Engineering, University of Science and Technology of China, Hefei 230026, China

[3] Beijing Research Institute of Mechanical & Electrical Technology, Beijing 100083, China; xuym@jds.ac.cn

* Correspondence: hitzhang@hit.edu.cn (Y.Z.); yanmufu@hit.edu.cn (M.Y.); xiacr@ustc.edu.cn (C.X.)

Received: 13 November 2020; Accepted: 10 December 2020; Published: 17 December 2020

Abstract: Surface exchange coefficient (k) and bulk diffusion coefficient (D) are important properties to evaluate the performance of mixed ionic-electronic conducting (MIEC) ceramic oxides for use in energy conversion devices, such as solid oxide fuel cells. The values of k and D are usually estimated by a non-linear curve fitting procedure based on electrical conductivity relaxation (ECR) measurement. However, the rate-limiting mechanism (or the availability of k and D) and the experimental imperfections (such as flush delay for gaseous composition change, τ_f) are not reflected explicitly in the time–domain ECR data, and the accuracy of k and D demands a careful sensitivity analysis of the fitting error. Here, the distribution of characteristic times (DCT) converted from time–domain ECR data is proposed to overcome the above challenges. It is demonstrated that, from the DCT spectrum, the rate-limiting mechanism and the effect of τ_f are easily recognized, and the values of k, D and τ_f can be determined conjunctly. A strong robustness of determination of k and D is verified using noise-containing ECR data. The DCT spectrum opens up a way towards visible and credible determination of kinetic parameters of MIEC ceramic oxides.

Keywords: distribution of characteristic times; electrical conductivity relaxation; surface exchange coefficient; bulk diffusion coefficient

---

1. Introduction

The high-temperature energy conversion technologies, such as solid oxide fuel cells (SOFCs) and solid oxide electrolysis cells (SOECs), are gaining worldwide research interest. One reason is that these technologies are conceptually highly efficient and ecofriendly, however they also face challenges in delivering high performance and longevity using low-cost materials [1–3]. Another reason is that they are a very good platform for the fundamental research of solid state ionics and electrochemistry [4–8]. One representative example is the development of mixed ionic-electronic conducting (MIEC) materials for use in the electrodes. The performance of MIECs are usually evaluated by the rate of the oxygen reduction/evolution reactions on the surface, and the diffusivity of oxygen ions in bulk, represented respectively by the surface exchange coefficient (k, cm·s^{-1}) and bulk diffusion coefficient (D, cm^2·s^{-1}) [9–15]. The measurements of k and D are usually based on ex-situ experiments, such as electrical conductivity relaxation (ECR) and isotope exchange depth profiling [12,13,15–17]. It has been shown that k and D determine the in-situ device-level performance, e.g., area specific resistance of electrodes [18]. In principle, the objective is to obtain high values of k and D, while considering the trade-off with longevity [2,19–21]. Researchers have developed atomic level descriptors that correlate to k and D for rational design of MIECs [3,22]. Therefore, the development and understanding of MIECs demands correct measurements of k and D.

Currently, the ECR method has been widely used to measure k and D, by fitting analytical solution to the relaxed conductivity data in response of a step change of atmosphere oxygen pressure. However, there is significant (orders of magnitude) discrepancy in the reported values of k and D for materials with the same nominal composition [23]. One reason is the incorporation of imperfections, such as the relaxation in initiating the driving force (e.g., the flush delay in change of oxygen pressure in atmosphere, τ_f), the nonlinearity if the sample is driven far away from equilibrium, the inhomogeneity of the sample and the instability at high temperatures [24]. Another reason for the unreliable chemical relaxation (CR) results derives from the trial-and-error approach. Reliable values of k and D can only be determined theoretically when the half thickness of the sample (L) close to a critical length ($L_c = D/k$), or in other words, the Biot number ($Bi = Lk/D$) being close to "1". Otherwise, the relaxation is governed either by bulk diffusion ($Bi \gg 1$), or by surface exchange ($Bi \ll 1$). Therefore, prior to measurements, the dimension of the sample should be designed properly based on the rational guess of k and D. This depends largely on experience. The Biot number (or the rate-limiting mechanism) is difficult to be identified from the ECR data, since the analytical solutions to the various rate-limiting mechanisms fit to the ECR data equally well. Efforts have been made to improve the reliability. For example, the flush delay can be shortened to a subsecond using small reactors [16]; The time–domain relaxation data can be converted into a frequency–domain electrochemical impedance spectroscopy (EIS), which is effective in capturing the flush delay and the rate-limiting step [25]; the confidence of the estimated k and D can be evaluated by a sensitivity analysis [26]. All the approaches are analogous to the EIS analysis methods based on equivalent circuit models. While there have been several emerging high-resolution methods for EIS analysis, such as the distribution of relaxation times (DRT) [27–29], there is a need to develop new methods independent of the trial-and-error approach for resolving of the CR kinetics and parameters.

By the existing ECR method, the analytical solutions to the ECR data are an infinite series of exponential terms represented by the characteristic times (CTs) and the corresponding pre-exponential factor (or strength) of each CT. Therein, the k and D are imbedded in the CTs and strengths in a non-linear, non-analytical manner. Therefore, the estimation of k and D using non-linear curve fitting procedure face challenges, although it is technically feasible. Recently, our team developed a new method namely distribution of characteristic times (DCT) to analyze the chemical relaxation in porous MIECs [30]. In that work, the time–domain ECR data is converted to a spectrum where the strength is plotted as a continuous function of logarithmic CT. For porous MIECs, the DCT spectrum was interpreted by a DCT model, and the values of k were estimated using the characteristics of the surface exchange-dominated peak in the DCT spectra. In principle, the DCT method is applicable to dense bulk MIEC bars/sheets/disks that are usually used in the literature. The DCT method provides a different manner to present the ECR data. As compared to the time–domain representation of ECR data, the DCT spectrum can visualize the CTs and their strengths without any preknowledge of the system. From the DCT spectrum, the intrinsic information of the ECR process can be visualized. In this work, the feasibility of estimation of k and D using the DCT spectrum reconstructed from the ECR data of dense bulk MIEC is demonstrated. It will be shown that the DCT method has a clear advantage over the time–domain analysis method.

2. The Theory of the Distribution of Characteristic Times

To deduce the DCT theory for dense bulk MIEC materials, it is necessary to review the analytical solution to the ECR data. For example, a dense MIEC sheet with a thickness $2L_x$ yields the solution [13],

$$\sigma(t) = 1 - \sum_{k=1}^{\infty} \frac{2Bi_x^2 \exp\left(-\beta_k^2 Dt/L_x{}^2\right)}{\beta_k^2\left(\beta_k^2 + Bi_x^2 + Bi_x\right)} \tag{1}$$

where the Biot numbers are given by,

$$Bi_x = L_x k/D \tag{2}$$

The dimensionless parameters β_k are the kth roots of the following equations,

$$\beta_k \tan \beta_k = Bi_x \tag{3}$$

The infinite series solution is obtained from the eigenfunctions for space and time by separation of the variables method. By inspection of Equation (1), we can see that this solution can be rewritten as a series of exponential functions of CTs (τ_i) and strengths (λ_i),

$$\sigma(t) = 1 - \sum_{i=1}^{\infty} \lambda_i \exp\left(-\frac{t}{\tau_i}\right) \tag{4}$$

where $\sum_{i=1}^{\infty} \lambda_i = 1$. It is noted that, if there is a relaxation in the change of atmosphere, say a flush delay of τ_f, Equation (4) is altered to [31],

$$\sigma(t) = 1 - \sum_{i=1}^{\infty} \lambda_i \frac{\tau_i}{\tau_i - \tau_f} \exp\left(-\frac{t}{\tau_i}\right) - \left(1 - \sum_{i=1}^{\infty} \lambda_i \frac{\tau_i}{\tau_i - \tau_f}\right) \exp\left(-\frac{t}{\tau_f}\right) \tag{5}$$

By treating τ_f as a new characteristic time, Equation (5) can be rewritten as,

$$\sigma(t) = 1 - \sum_{i=1}^{\infty} \chi_i \exp(-t/\tau_i) \tag{6}$$

where $\sum_{i=1}^{\infty} \chi_i = 1$. Obviously, if the change of atmosphere is instantaneous ($\tau_f = 0$), we have $\chi_i > 0$, else if there is a flush delay subject to $\tau_f << \tau_1$, we have $\chi_i \leq 0$ for $\tau_i \leq \tau_f$, and $\chi_i > 0$ for $\tau_i > \tau_f$. It is natural to see that the right-hand-side of Equation (6) can be generally represented by an integral,

$$\sigma(t) = 1 - \int_{-\infty}^{+\infty} [\chi \exp(-t/\tau)] d\log_{10}\tau \tag{7}$$

subject to,

$$\int_{-\infty}^{+\infty} \chi d\log_{10}\tau = 1 \tag{8}$$

Equation (7) suggests two manners to represent the ECR data. The left-hand-side of Equation (7) corresponds to a time–domain representation, such as a plot of t versus σ. By this plot, it is easy to present the experimental ECR data, however, it is hard to see the rate-limiting step of the ECR experiment that is vital to the determination of k and D. In contrast, the right-hand-side of Equation (7) suggests a DCT representation, such as a plot of $\log_{10}\tau$ versus χ. The DCT function $\chi(\log_{10}\tau)$ can be reconstructed from the experimental $\sigma(t)$ data, by converting Equations (7) and (8) to a standard form of bound constrained quadratic programming problem based on a Tikhonov regularization [30]. A plot of $\log_{10}\tau$ versus χ represents a new type of spectrum, showing the kinetic details that are hardly detectable by a time–domain representation of the ECR data.

3. Results and Discussion

A systematic study for the feasibility of resolving the values of k and D from ECR data was verified based on artificial ECR experiments. The ECR data was generated using the analytical solution of the MIEC sheet, for which the thickness ($2L_x$) is much smaller than the other dimensions. k and D are given as 1×10^{-4} cm·s^{-1}, and 1×10^{-5} cm^2·s^{-1}, respectively. L_x serves as a variable parameter to simulate the effect of Biot number ($Bi \equiv L_x k/D$). The analytical solution of DCT, represented by a series of CTs and the corresponding strengths (τ_i, A_i) was obtained directly from the analytical ECR solution. On the other hand, the DCT represented by the $\log_{10}\tau$–χ relationship was reconstructed using the synthetic ECR data. The following contents first show the feasibility and merits of DCT spectroscopy in visualizing the rate-limiting step and the flush delay. Then, the robustness against noise and the accuracy in determining the values of k and D were demonstrated.

Figure 1 shows the comparison between the various representations of the synthetic ECR data with different Biot numbers. From the time–domain representation of the ECR data (Figure 1a,c,e) under Biot numbers of 0.01 (surface-exchange limited), 1 (surface-exchange & bulk-diffusion co-limited) and 100 (bulk-diffusion limited), it is hard to distinguish the rate-limiting mechanisms. From the reconstructed DCT spectra (the black curves in Figure 1b,d,f), it is shown clearly that the DCT shows a single peak for $Bi = 0.01$. For $Bi = 1$ and 100, the second-order peak appears in the DCT, with the strength increasing with Bi. It is noted that, for $Bi = 100$, the higher-order peaks appeared, which cannot be reconstructed properly by the DCT. These incorrect peaks may be derived from the Tikhonov regularization, since their strengths are negligible as compared to the first-order peaks. The similar problem is also encountered in reconstructing the DRT from the EIS spectrum by Tikhonov regularization [32]. Although the third and higher order peaks may be miscalculated, the first two major peaks with significant strength play a dominant role in visualizing the rate-limiting step. It is shown by the τ_i–A_i plot that the first- and second-order peaks of the reconstructed DCT agree well with the theoretical solutions (the orange stems in Figure 1b,d,f). Therefore, the reconstructed DCT spectrum can be used to identify the rate-limiting mechanism. To provide a whole picture that how the major peak(s) of DCT evolves with rate-limiting regimes, a contour plot of the reconstructed DCT in the logarithmic Bi–τ axis system is presented in Figure 2a.

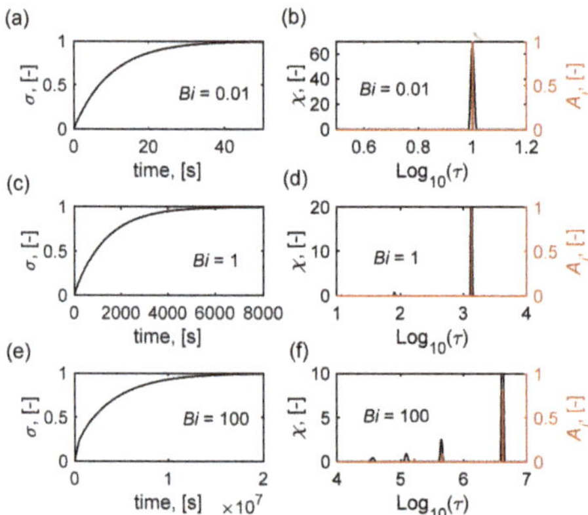

Figure 1. Comparison of the time domain representation of the electrical conductivity relaxation (ECR) data and the reconstructed distribution of characteristic times (DCT) spectra. (**a,c,e**) The time–domain representation of ECR data calculated from analytical solution of the mixed ionic-electronic conducting (MIEC) sheet with $k = 10^{-4}$ cm·s^{-1}, $D = 10^{-5}$ cm^2·s^{-1} and a Biot number of 0.01 (**a**), 1 (**c**) and 100 (**e**). (**b,d,f**) The DCT spectra calculated from (**a,c,e**) and the corresponding analytical solutions of DCT.

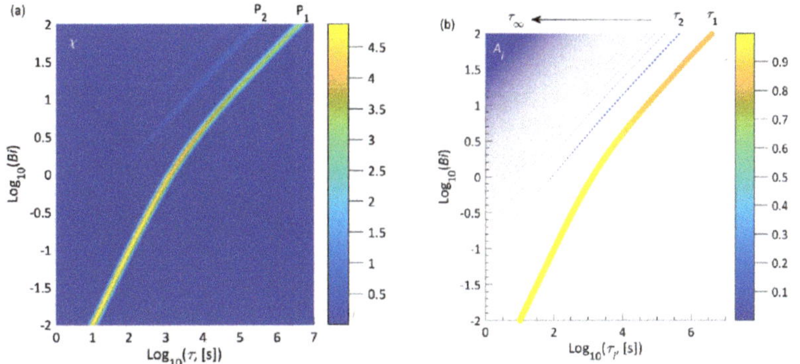

Figure 2. The reconstructed DCT spectra (**a**) and the analytical solution of DCT (**b**) of MIEC sheets at various Biot numbers, with $k = 10^{-4}$ cm·s^{-1} and $D = 10^{-5}$ cm^2·s^{-1}. The strength of characteristic time (CT) scales with the color (**a**,**b**) and the size of scatter (**b**).

It is shown that, for the surface-exchange-limited kinetics (e.g., $Bi < 10^{-1}$), the DCT presents only the first-order peak (P1). For the bulk-diffusion-limited kinetics (e.g., $Bi > 10$), the second-order peak (P2) appeared. In this condition, the ratio of the CTs of P1 and P2 was $\tau_{P1}/\tau_{P2} = 9$. For the colimited kinetics (e.g., $10^{-1} < Bi < 10$), the ratio was subject to $\tau_{P1}/\tau_{P2} > 9$. These features were verified by comparing with a scatter plot of the analytical DCT shown in Figure 2b. Therefore, the rate-limiting mechanism, which was hardly revealed from the time–domain representation of the ECR data, can be visualized explicitly in the DCT spectrum.

Then, we turned to the issue of determining the values of k and D from the DCT spectrum. Mathematically, k and D can be determined using two quantitative characteristics of the DCT, for example the CT (τ_{P1}) and the strength (S_{P1}) of P1. In the surface-exchange limited regime, the DCT shows one peak (P1). In this case, only the value of k can be determined, given by,

$$k = L_x/\tau_{P1} \tag{9}$$

In the bulk-diffusion limited regime, the DCT shows two or more peaks subject to $\tau_{P1}/\tau_{P2} = 9$. In this case, only the value of D can be determined, given by,

$$D = 4L_x{}^2/\left(\pi^2\tau_{P1}\right) \tag{10}$$

In the co-limited regime, both the values of k and D are obtainable. In this case, both S_{P1} and τ_{P1} are needed, following

$$\tau_{P1} = \frac{L_x{}^2}{D\alpha_1{}^2} \tag{11}$$

$$S_{P1} = \frac{2Bi^2}{\alpha_1{}^2(\alpha_1{}^2 + Bi^2 + Bi)} \tag{12}$$

$$Bi \equiv \frac{L_x k}{D} = \alpha_1 \tan \alpha_1 \tag{13}$$

The first-order root of Equation (13), α_1 approaches to 0, and to $\pi/2$ for the surface-exchange-limited and the bulk-diffusion-limited kinetics, respectively. Obviously, the accuracy of S_{P1} and τ_{P1} revealed by the DCT spectrum is vital to the determination of k and/or D. Figure 3 shows that the values of S_{P1} and τ_{P1}, resolved from the DCT spectra in Figure 2a, fit excellently to the analytical values (A_1 and τ_1).

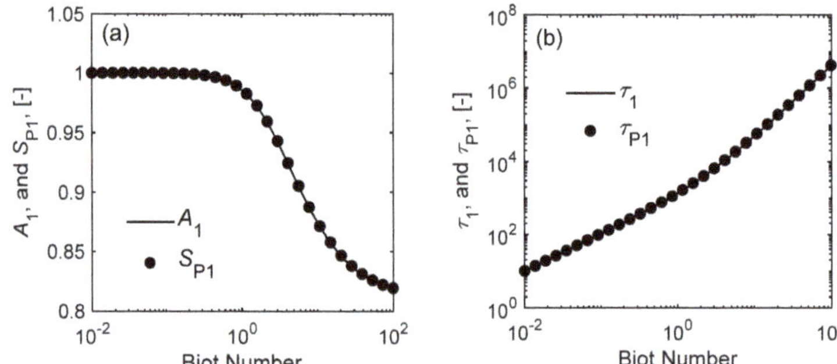

Figure 3. (**a**) The major peak strength of the reconstructed DCT (the integral area of the peak P1 in Figure 2a, S_{P1}) and the analytical strength of τ_1 (A_1), and (**b**) the CTs of the major peak in the reconstructed DCT (τ_{P1}) and the analytical values of τ_1 of MIEC sheets with $k = 10^{-4}$ cm·s^{-1} and $D = 10^{-5}$ cm^2·s^{-1}, as a function of the Biot number.

The above discussion shows that it is theoretically feasible to determine the values of k and D from the DCT spectrum. However, in practice, non-ideal factors are usually imbedded in the ECR data that can influence the quality of DCT spectrum, therefore the determination of k and D. One major factor is the flush delay (τ_f), denoting the relaxation time of the gaseous composition change. By considering the effect of τ_f, Equation (12) is revised to be,

$$S_{P1} \frac{\tau_{P1} - \tau_f}{\tau_{P1}} = \frac{2Bi^2}{\alpha_1{}^2(\alpha_1{}^2 + Bi^2 + Bi)} \tag{14}$$

Therefore, the determination of k and D demands the correct estimation of S_{P1}, τ_{P1} and τ_f. It is hard to detect the effect of τ_f on the ECR data from the time–domain representation. However, as shown by Equation (5), the strengths of the CTs below τ_f are theoretically negative values. Therefore, the DCT spectrum is expected to be a visible manner identifying the flush delay. Figure 4 demonstrates an example that how a rational DCT spectrum and a correct value of τ_f can be determined from a flush-delay-imbedded ECR data.

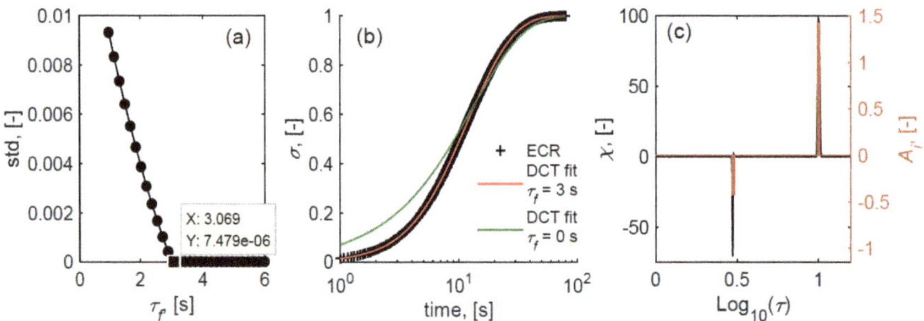

Figure 4. Reconstruction of DCT from the analytical ECR solution of an MIEC sheet with a flush delay $\tau_f = 3$ s, $k = 10^{-4}$ cm·s^{-1}, $D = 10^{-5}$ cm^2·s^{-1} and a Biot number of 0.01. (**a**) The standard deviation (std) between DCT fitting and the original ECR data as a function of τ_f for use in DCT reconstruction. (**b**) The comparison between the original ECR data and the DCT fitting with different values of τ_f. (**c**) The reconstructed DCT spectrum (the left axis), and the analytical DCT (the left axis).

The ECR data was generated using Equation (5) with $\tau_f = 3$ s, $k = 10^{-4}$ cm·s^{-1}, $D = 10^{-5}$ cm^2·s^{-1} and $Bi = 0.01$. For reconstruction of DCT, a guess of τ_f must be given, since DCT is constricted to be negative/positive for CTs below/above τ_f. To determine the value of τ_f, a parametric sweep study on τ_f

is performed to obtain the evolution of standard deviation between the original ECR and the ECR determined by the reconstructed DCT, as shown by Figure 4a. It is shown that the evolution of standard deviation with τ_f shows an L-shaped curve. The corner point of the L-shaped curve can be used as a good estimator of τ_f, which was estimated to be 3.07 s for this case (very close to the theoretical value of 3 s). Figure 4b shows the comparison of the original ECR and the DCT-fitted ECR, showing that the ECR could be perfectly fitted by the DCT when a correct value of τ_f is given. The corresponding DCT spectrum, as shown by Figure 4c, shows a negative peak at the CT of 3 s. To check the effectiveness of the procedure, a case study of τ_f with different values are shown in Figure 5, showing that the values of τ_f can be estimated correctly by the L-shaped curve of standard deviation.

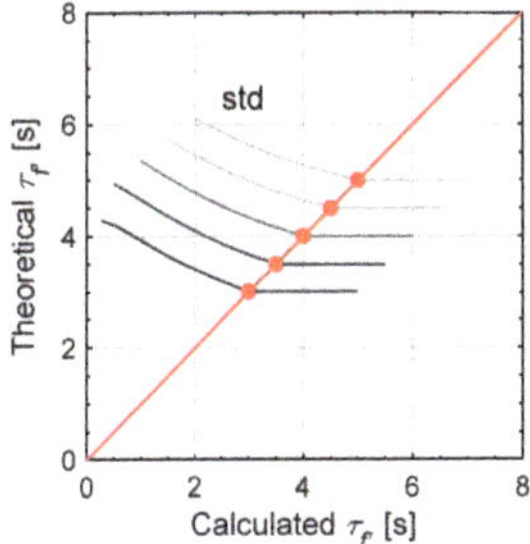

Figure 5. Determination of τ_f for the various flush-delay-containing ECR data by the evolution of standard deviation (std) with a sweep of τ_f. The red dots represent the coordinates of the calculated τ_f and the theoretical τ_f determined as the point where the std does not decrease obviously with τ_f. The 45° line is for eye-guide.

Another factor that influences the quality of ECR data is the noisy perturbations. Therefore, the robustness of DCT reconstruction using noise-containing ECR data must be verified for practical usage. Based on the flush-delay-imbedded ECR, the effect of noise on the DCT reconstruction was studied by adding a random Gaussian noise with a standard deviation of 1% in the ECR data in Figure 4. Figure 6a shows the L-shaped curves of standard deviation for three-times repeats of parametric sweep on τ_f, showing that a correct value of 3 s can be estimated. To check if the correct values of S_{P1} and τ_{P1} can be resolved from the DCT, the noise-containing ECR data with different Biot numbers were simulated. Figure 6b,c shows that the values of S_{P1} and τ_{P1} resolved from the reconstructed DCT agreed nicely to the analytical values.

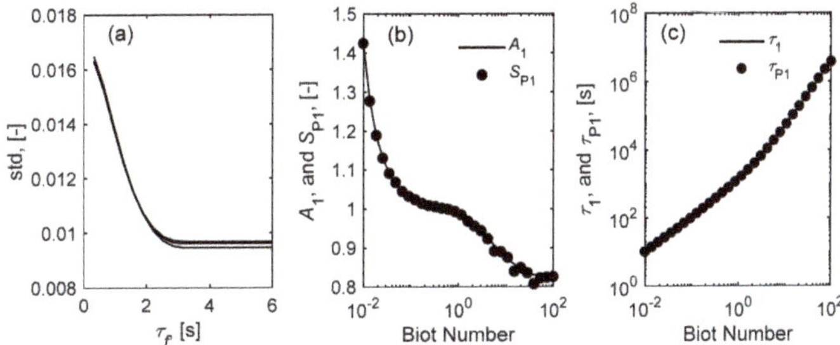

Figure 6. Determination of the flush delay ((**a**) the three lines represent the results of three ECR datasets with the same theoretical solution imbedded with 1% standard deviation level random Gaussian noise), (**b**) the major peak strength of the reconstructed DCT (S_{P1}) and the analytical strength of τ_1 (A_1) and (**c**) the CTs of the major peak in the reconstructed DCT (τ_{P1}) and the analytical values of τ_1 for noise-containing ECR data of MIEC sheet with a flush delay $\tau_f = 3$ s, $k = 10^{-4}$ cm·s^{-1} and $D = 10^{-5}$ cm^2·s^{-1}. A random Gaussian noise with a 1% standard deviation level is added in the ECR data at each Biot number.

Finally, we demonstrated the capability of determining the values of k and D using the DCT spectrum reconstructed from noise-containing ECR data with different levels of flush delay ($\tau_f/\tau_1 = 0$, 0.2, 0.4) and Biot numbers ($Bi = 10^{-2} \sim 10^2$), as shown in Figure 7a–c.

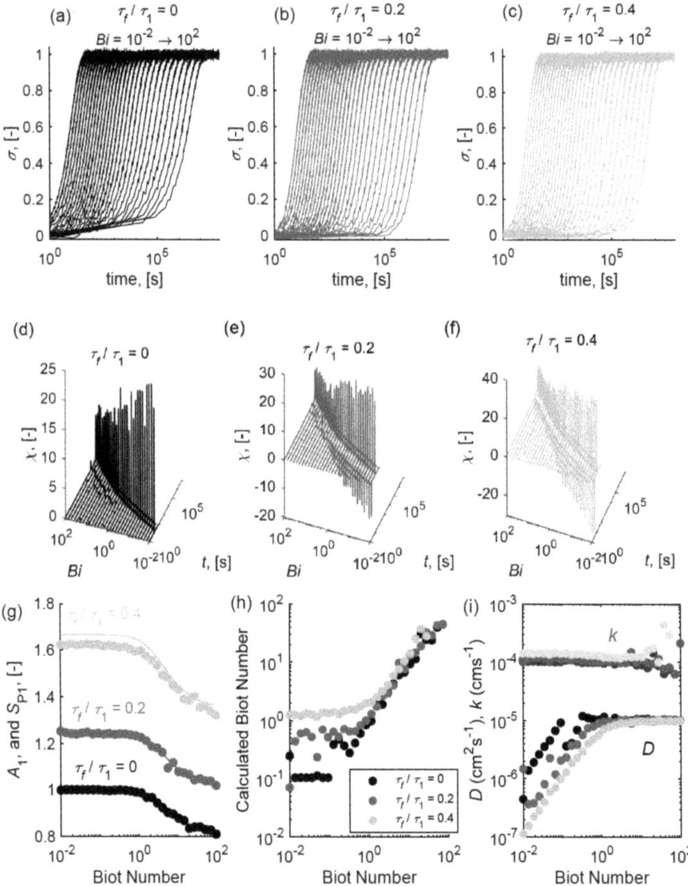

Figure 7. Determination of k and D from reconstruction of DCT. (**a–c**) The synthesized ECR data for the MIEC sheets with $k = 10^{-4}$ cm·s^{-1}, $D = 10^{-5}$ cm^2·s^{-1}, std = 1%, Bi = $10^{-2} \sim 10^2$ and $\tau_f/\tau_1 = 0$ (a), $\tau_f/\tau_1 = 0.2$ (**b**) and $\tau_f/\tau_1 = 0.4$ (c). (**d–f**) The reconstructed DCT from the synthesized noise-containing ECR datasets with $\tau_f/\tau_1 = 0$ (**d**), $\tau_f/\tau_1 = 0.2$ (**e**) and $\tau_f/\tau_1 = 0.4$ (**f**). (**g–i**) The evolution of the major peak strength of the reconstructed DCT (S_{P1}) and the analytical strength of τ_1 (A_1) (**g**), the calculated Biot number (**h**) and the calculated D and k (**i**) as a function of Biot number from the reconstructed DCT spectra with different levels of τ_f.

It is shown that the time–domain ECR curves present a similar shape, from which it is hard to identify the flush delay. The reconstructed DCT spectra are shown in Figure 7d–f, showing that the flush delay can be explicitly represented by the negative peak. For $\tau_f/\tau_1 = 0$ (Figure 7d), the rate-limiting mechanism can be captured by the DCT spectra. For $\tau_f/\tau_1 = 0.2$ and 0.4 (Figure 7e,f), it is noted that, for bulk-diffusion-limited kinetics, the second-order peak was not reconstructed properly and pseudo small peaks around τ_f appeared. This resulted from the high level of flush delay that was higher than the CT of the second-order peak ($\tau_2/\tau_1 = 1/9$), so that the second-order peak was altered to be negative. Another reason is the strong interaction between τ_f and τ_2 that cannot be properly decoupled by the present Tikhonov regularization method. The same challenge is also encountered in the reconstruction of DRT from impedance spectroscopy. That is, the small polarization resistance process is hard to be decoupled by the DRT from the high resistance process with a similar relaxation time [32]. Pseudo small DRT peaks are hard to be avoided by the present methods, such as Tikhonov regularization, and Fourier transform [24]. In analogous to DRT, the reconstruction of DCT requires high quality ECR data. In practice, the flush delay is typically several seconds, which is usually much smaller than τ_1 (e.g., thousands of seconds for dense bulk MIECs [11,17,33,34]). If the flush delay is significant so that the second-order peak interplays strongly with the flush delay peak ($\tau_2 \approx \tau_f$), the DCT spectrum can still serve as a tool for identifying the flush delay. For determination of k and D, the first-order peak and the flush delay played a crucial role. Figure 7d–f shows that τ_{P1} and τ_f could be correctly estimated. Figure 7g shows that the value of S_{P1} could be properly estimated for the various flush delays and Biot numbers. It is shown that the value of S_{P1} increased with the increase of τ_f/τ_{P1}. From Equation (14), we can see that S_{P1} is proportional to τ_f/τ_{P1} at a fixed Biot number. Equation (14) also indicates that, combining with Equation (13), the value of the Biot number can be estimated using the values of τ_f, τ_{P1} and S_{P1}. Figure 7h shows the calculated Biot numbers as a function of the theoretical Biot number for the different levels of flush delay. It is shown that the Biot number could be determined properly above a theoretical value of "1". The calculated Biot number was overestimated if the value was below "1". This overestimation resulted from the low estimation of S_{P1} at small Biot numbers, as shown by the case $\tau_f/\tau_1 = 0.4$ in Figure 7g. According to the estimated Biot number, the values of k and D could be estimated using Equations (11) and (13). It is shown in Figure 7i that the value of k could be properly estimated under $Bi < 10$, while the value of D could be properly estimated under $Bi > 1$. It is noted that, for the case $\tau_f/\tau_{P1} = 0$, the value of D is obtainable under $Bi > 0.1$, for which the Biot number can be properly estimated (Figure 7h). Therefore, the DCT spectrum demonstrated the same capability with the time–domain analysis method in determining k and D when the flush delay is negligible [26]. For the case that flush delay was significant, the DCT spectrum was capable of determining the value of k and D when the estimated Biot number was below "10" and above "1", respectively. More importantly, the value of τ_f could be determined explicitly. The conjunct determination of k, D and τ_f was not demonstrated by the existing time–domain analysis methods. The DCT spectrum provides a visual, high-resolution and reliable manner for revealing chemical relaxation processes and resolving the kinetic parameters.

4. Conclusions

In summary, the DCT spectrum converted from ECR data was demonstrated to be a visual, high-resolution and robust tool for revealing the chemical relaxation kinetics in the dense bulk MIEC materials. As compared to the time–domain representation of ECR data, the DCT spectrum was capable of visualizing the rate-limiting mechanism and the flush delay. The values of surface exchange and bulk diffusion coefficients and flush delay could be determined conjunctly using the characteristics of the first-order peak and the flush delay peak of the DCT spectrum. The robustness against noise of DCT reconstruction was verified to be sufficient for practical usage. The DCT method was generally applicable to other types of chemical relaxation measurements for revealing rate-limiting steps and determining kinetic parameters, such as carburizing and nitriding of steels.

Author Contributions: Data curation, F.Y.; Formal analysis, F.Y.; Funding acquisition, Y.Z., M.Y., and Y.X.; Investigation, F.Y., Y.W., Y.Y., L.Z., H.H., Z.T., and C.X.; Methodology, Y.Z. All authors have read and agreed to the published version of the manuscript.

Funding: This research was funded by the National Key Research and Development Program of China (2018YFB2001901) and National Natural Science Foundation of China (21673062, 52,073,072 and 51972298), and the Key Research and Development Program of Sichuan Province (2020YFSY0026).

Conflicts of Interest: There is no conflict of interest.

References

1. Zhao, C.; Li, Y.; Zhang, W.; Zheng, Y.; Lou, X.; Yu, B.; Chen, J.; Chen, Y.; Liu, M.; Wang, J. Heterointerface engineering for enhancing the electrochemical performance of solid oxide cells. *Energy Environ. Sci.* **2020**, *13*, 53–85. [CrossRef]
2. Zhang, Y.; Knibbe, R.; Sunarso, J.; Zhong, Y.; Zhou, W.; Shao, Z.; Zhu, Z. recent progress on advanced materials for solid-oxide fuel cells operating below 500 °C. *Adv. Mater.* **2017**, *29*, 1700132. [CrossRef] [PubMed]
3. Zheng, Y.; Wang, J.; Yu, B.; Zhang, W.; Chen, J.; Qiao, J.; Zhang, L. A review of high temperature co-electrolysis of H_2O and CO_2 to produce sustainable fuels using solid oxide electrolysis cells (SOECs): Advanced materials and technology. *Chem. Soc. Rev.* **2017**, *46*, 1427–1463. [CrossRef] [PubMed]
4. Chen, D.; Guan, Z.; Zhang, D.; Trotochaud, L.; Crumlin, E.J.; Nemsak, S.; Bluhm, H.; Tuller, H.L.; Chueh, W.C. Constructing a pathway for mixed ion and electron transfer reactions for O_2 incorporation in $Pr_{0.1}Ce_{0.9}O_{2-x}$. *Nat. Catal.* **2020**, *3*, 116–124. [CrossRef]
5. Poetzsch, D.; Merkle, R.; Maier, J. Stoichiometry variation in materials with three mobile carriers-thermodynamics and transport kinetics exemplified for protons, oxygen vacancies, and holes. *Adv. Funct. Mater.* **2015**, *25*, 1542–1557. [CrossRef]
6. Zhao, F.; Virkar, A. Dependence of polarization in anode-supported solid oxide fuel cells on various cell parameters. *J. Power Sources* **2005**, *141*, 79–95. [CrossRef]
7. Irvine, J.T.; Neagu, D.; Verbraeken, M.C.; Chatzichristodoulou, C.; Graves, C.; Mogensen, M.B. Evolution of the electrochemical interface in high-temperature fuel cells and electrolysers. *Nat. Energy* **2016**, *1*, 15014. [CrossRef]
8. Gregori, G.; Merkle, R.; Maier, J. Ion conduction and redistribution at grain boundaries in oxide systems. *Prog. Mater. Sci.* **2017**, *89*, 252–305. [CrossRef]
9. Bucher, E.; Egger, A.; Ried, P.; Sitte, W.; Holtappels, P. Oxygen nonstoichiometry and exchange kinetics of $Ba_{0.5}Sr_{0.5}Co_{0.8}Fe_{0.2}O_{3-\delta}$. *Solid State Ion.* **2008**, *179*, 1032–1035. [CrossRef]
10. Sunarso, J.; Baumann, S.; Serra, J.M.; Meulenberg, W.A.; Liu, S.; Lin, Y.S.; Da Costa, J.C.D. Mixed ionic–electronic conducting (MIEC) ceramic-based membranes for oxygen separation. *J. Membr. Sci.* **2008**, *320*, 13–41. [CrossRef]
11. Ko, H.-D.; Lin, C.-C. Oxygen Diffusivities and Surface Exchange Coefficients in Porous Mullite/Zirconia Composites Measured by the Conductivity Relaxation Method. *J. Am. Ceram. Soc.* **2010**, *93*, 176–183. [CrossRef]
12. Kilner, J. Surface exchange of oxygen in mixed conducting perovskite oxides. *Solid State Ion.* **1996**, *86*, 703–709. [CrossRef]
13. Yasuda, I.; Hishinuma, M. Electrical conductivity and chemical diffusion coefficient of strontium-doped lanthanum manganites. *J. Solid State Chem.* **1996**, *123*, 382–390. [CrossRef]
14. De Souza, R.A.; Kilner, J.A. Oxygen transport in $La_{1-x}Sr_xMn_{1-y}Co_yO_{3\pm\delta}$ perovskites: Part II. Oxygen surface exchange. *Solid State Ion.* **1999**, *126*, 153–161. [CrossRef]
15. Lane, J. Measuring oxygen diffusion and oxygen surface exchange by conductivity relaxation. *Solid State Ion.* **2000**, *136*, 997–1001. [CrossRef]
16. Ganeshananthan, R.; Virkar, A.V. Measurement of surface exchange coefficient on porous $La_{0.6}Sr_{0.4}CoO_{3-\delta}$ samples by conductivity relaxation. *J. Electrochem. Soc.* **2005**, *125*, A1620–A1628. [CrossRef]
17. Zhang, L.; Liu, Y.; Zhang, Y.; Xiao, G.; Chen, F.; Xia, C. Enhancement in surface exchange coefficient and electrochemical performance of $Sr_2Fe_{1.5}Mo_{0.5}O_6$ electrodes by $Ce_{0.8}Sm_{0.2}O_{1.9}$ nanoparticles. *Electrochem. Commun.* **2011**, *13*, 711–713. [CrossRef]
18. Adler, S.B. Factors Governing Oxygen Reduction in Solid Oxide Fuel Cell Cathodes. *Chem. Rev.* **2004**, *104*, 4791–4844. [CrossRef]
19. Chen, Y.; Yoo, S.; Choi, Y.; Kim, J.H.; Ding, Y.; Pei, K.; Murphy, R.; Zhang, Y.; Zhao, B.; Zhang, W.; et al. A highly active, CO_2-tolerant electrode for the oxygen reduction reaction. *Energy Environ. Sci.* **2018**, *11*, 2458–2466. [CrossRef]
20. Chen, Y.; Choi, Y.; Yoo, S.; Ding, Y.; Yan, R.; Pei, K.; Qu, C.; Zhang, L.; Chang, I.; Zhao, B.; et al. A highly efficient multi-phase catalyst dramatically enhances the rate of oxygen reduction. *Joule* **2018**, *2*, 938–949. [CrossRef]
21. Chen, Y.; Bu, Y.; Zhang, Y.; Yan, R.; Ding, D.; Zhao, B.; Yoo, S.; Dang, D.; Hu, R.; Yang, C.; et al. A highly efficient and robust nanofiber cathode for solid oxide fuel cells. *Adv. Energy Mater.* **2016**, *7*, 1601890. [CrossRef]
22. Lee, Y.-L.; Kleis, J.; Rossmeisl, J.; Shao-Horn, Y.; Morgan, D. Prediction of solid oxide fuel cell cathode activity with first-principles descriptors. *Energy Environ. Sci.* **2011**, *4*, 3966–3970. [CrossRef]

23. Hu, B.; Xia, C. Factors influencing the measured surface reaction kinetics parameters. *Asia Pac. J. Chem. Eng.* **2016**, *11*, 327–337. [CrossRef]

24. Ciucci, F. Modeling electrochemical impedance spectroscopy. *Curr. Opin. Electrochem.* **2019**, *13*, 132–139. [CrossRef]

25. Boukamp, B.; Otter, M.; Bouwmeester, H. Transport processes in mixed conducting oxides: Combining time domain experiments and frequency domain analysis. *J. Solid State Electrochem.* **2004**, *8*, 592–598. [CrossRef]

26. Ciucci, F. Electrical conductivity relaxation measurements: Statistical investigations using sensitivity analysis, optimal experimental design and ECRTOOLS. *Solid State Ion.* **2013**, *239*, 28–40. [CrossRef]

27. Zhang, Y.; Chen, Y.; Chen, F. In-situ quantification of solid oxide fuel cell electrode microstructure by electrochemical impedance spectroscopy. *J. Power Sources* **2015**, *277*, 277–285. [CrossRef]

28. Zhang, Y.; Chen, Y.; Yan, M.F.; Chen, F. Reconstruction of relaxation time distribution from linear electrochemical impedance spectroscopy. *J. Power Sources* **2015**, *283*, 464–477. [CrossRef]

29. Xia, J.; Wang, C.; Wang, X.; Bi, L.; Zhang, Y. A perspective on DRT applications for the analysis of solid oxide cell electrodes. *Electrochim. Acta* **2020**, *349*, 136328. [CrossRef]

30. Zhang, Y.; Yan, F.; Hu, B.; Xia, C.; Yan, M. Chemical relaxation in porous ionic–electronic conducting materials represented by the distribution of characteristic times. *J. Mater. Chem. A* **2020**, *8*, 17442–17448. [CrossRef]

31. Otter, M.W.D.; Bouwmeester, H.J.; Boukamp, B.A.; Verweij, H. Reactor flush time correction in relaxation experiments. *J. Electrochem. Soc.* **2001**, *148*, J1–J6. [CrossRef]

32. Zhang, Y.; Chen, Y.; Li, M.; Yan, M.F.; Ni, M.; Xia, C. A high-precision approach to reconstruct distribution of relaxation times from electrochemical impedance spectroscopy. *J. Power Sources* **2016**, *308*, 1–6. [CrossRef]

33. Yeh, T.C.; Routbort, J.L.; Mason, T.O. Oxygen transport and surface exchange properties of $Sr_{0.5}Sm_{0.5}CoO_{3-\delta}$. *Solid State Ion.* **2013**, *232*, 138–143. [CrossRef]

34. Gopal, C.B.; Haile, S.M. An electrical conductivity relaxation study of oxygen transport in samarium doped ceria. *J. Mater. Chem. A* **2014**, *2*, 2405–2417. [CrossRef]

Publisher's Note: MDPI stays neutral with regard to jurisdictional claims in published maps and institutional affiliations.

© 2020 by the authors. Licensee MDPI, Basel, Switzerland. This article is an open access article distributed under the terms and conditions of the Creative Commons Attribution (CC BY) license (http://creativecommons.org/licenses/by/4.0/).

Article

A Novel Decarburizing-Nitriding Treatment of Carburized/through-Hardened Bearing Steel towards Enhanced Nitriding Kinetics and Microstructure Refinement

Fuyao Yan [1], Jiawei Yao [1], Baofeng Chen [1], Ying Yang [1], Yueming Xu [2], Mufu Yan [1,*] and Yanxiang Zhang [1,*]

[1] School of Materials Science and Engineering, Harbin Institute of Technology, Harbin 150001, China; fyan@hit.edu.cn (F.Y.); yjw0573@sina.com (J.Y.); baofengchen0313@163.com (B.C.); yangying960720@163.com (Y.Y.)

[2] Beijing Research Institute of Mechanical & Electrical Technology, Beijing 100083, China; yaojiawei0601@163.com

* Correspondence: yanmufu@hit.edu.cn (M.Y.); hitzhang@hit.edu.cn (Y.Z.); Tel.: +86-451-86418617 (M.Y.)

Abstract: Decarburization is generally avoided as it is reckoned to be a process detrimental to material surface properties. Based on the idea of duplex surface engineering, i.e., nitriding the case-hardened or through-hardened bearing steels for enhanced surface performance, this work deliberately applied decarburization prior to plasma nitriding to cancel the softening effect of decarburizing with nitriding and at the same time to significantly promote the nitriding kinetics. To manifest the applicability of this innovative duplex process, low-carbon M50NiL and high-carbon M50 bearing steels were adopted in this work. The influence of decarburization on microstructures and growth kinetics of the nitrided layer over the decarburized layer is investigated. The metallographic analysis of the nitrided layer thickness indicates that high carbon content can hinder the growth of the nitrided layer, but if a short decarburization is applied prior to nitriding, the thickness of the nitrided layer can be significantly promoted. The analysis of nitriding kinetics shows that decarburization reduces the activation energy for nitrogen diffusion and enhances nitrogen diffusivity. Moreover, the effect of decarburization in air can promote surface microstructure refinement via spinodal decomposition during plasma nitriding.

Keywords: duplex surface engineering; decarburization; carburization; plasma nitriding; nitriding kinetics

check for **updates**

Citation: Yan, F.; Yao, J.; Chen, B.; Yang, Y.; Xu, Y.; Yan, M.; Zhang, Y. A Novel Decarburizing-Nitriding Treatment of Carburized/through-Hardened Bearing Steel towards Enhanced Nitriding Kinetics and Microstructure Refinement. *Coatings* **2021**, *11*, 112. https://doi.org/10.3390/coatings11020112

Academic Editor: Emerson Coy

Received: 1 December 2020

Accepted: 13 January 2021

Published: 20 January 2021

Publisher's Note: MDPI stays neutral with regard to jurisdictional claims in published maps and institutional affiliations.

Copyright: © 2021 by the authors. Licensee MDPI, Basel, Switzerland. This article is an open access article distributed under the terms and conditions of the Creative Commons Attribution (CC BY) license (https://creativecommons.org/licenses/by/4.0/).

1. Introduction

M50 steel and its variant M50NiL steels are used as aero-engines bearings for high-temperature applications [1]. To maintain structural integrity in a more severe service environment, excellent combined surface properties, such as hardness, wear resistance and rolling contact fatigue resistance, are required for bearings. Traditional carburized case or through-hardened steel with secondary hardening effect is typically less than 65 HRC (~860 HV), with limited resistance in wear or other surface damages. On the other hand, nitrided layers can exhibit hardness greater than 1000 HV, but the typical case thickness is only 1/10 of the carburized case, and with a sharp decreasing hardness profile, which is not conductive to load-bearing capacities and fatigue resistance. Combination of carburizing/through-hardening and nitriding therefore becomes the typical duplex hardening techniques.

For M50 steel with ~0.8 wt.% carbon, duplex hardening achieved by applying a thin nitride layer over the secondary-hardened matrix [2], has been applied to further improve the surface properties. Ooi et al. reported that duplex hardening treatment can increase surface compression stress and therefore improve rolling contact fatigue resistance of the M50 components [3]. Rhoads et al. improved the surface hardness of

M50 to HV 1000~1250 and prolonged the rolling contact fatigue life by 8~10 times with duplex hardening [4]. Streit et al. reported that duplex-hardened M50 bearings had lower spall propagation rates over conventional counterparts in low or boundary lubrication conditions and contaminated lubrication conditions due to the high surface hardness [5]. For steels with reduced carbon content, duplex hardening can be achieved for surface by nitriding the pre-carburized case [6]. Carburizing is to obtain a thick hardening case to resist the Hertz stresses during loading, while nitriding can further increase the outmost surface hardness and generate more compressive residual stresses. For example, Davies et al. showed that duplex hardening can improve the axial and rolling contact fatigue life of helicopter gears made of Vasco X-2M steel [7].

However, duplex hardening processes are generally time-consuming due to the carbon/nitrogen trapping model [8–11]. Bloyce et al. applied duplex hardening to M50NiL steel and produced a nitrided case thickness of 100 μm with surface hardness over 1000 HV after nitriding at 400 °C for 60 h [12]. Egert et al. observed decarburization of steel surface during plasma nitriding and found that the removal of surface carbon can facilitate the diffusion of nitrogen into the bulk [13]. In general, decarburization during heat treatment is well-limited, since it can greatly decrease hardness and wear resistance of the surface [14]. However, this work is intended to deliberately apply decarburization prior to plasma nitriding, with the aim of cancelling the softening effect of decarburization with nitrided case, and at the same time aiming to promote the nitriding kinetics. In this work, low-carbon M50NiL steel and high-carbon M50 steel that represent two grades of steels were adopted as examples. The influence of decarburization on microstructures and growth kinetics of the nitrided layer is investigated in detail.

2. Materials and Methods

2.1. Materials

The materials used in this work were high-carbon M50 steel with the nominal composition Fe-0.85C-4Cr-4Mo-1V-0.15Mn-0.1Si (in wt.%) and low-carbon M50NiL steel with the nominal composition Fe-0.13C-4.1Cr-3.4Ni-4.2Mo-1.2V-0.13Mn-0.18Si (in wt.%). The steel bars were cut into 13 mm × 13 mm × 5 mm pieces for thermochemical treatment and microstructural characterizations. All specimens were mechanically ground by 240- and 800-grit sandpapers and ultrasonically cleaned in ethanol before thermochemical treatment.

2.2. Thermochemical Surface Treatment

The thermochemical surface treating process (taken as "duplex process" in the following for short) for M50NiL and M50 are schematically shown in Figure 1, in which a decarburizing step was added prior to plasma nitriding. Figure 1a shows the "duplex process" for M50NiL, which followed the sequence of gas carburizing, decarburizing and plasma nitriding. In Figure 1b, the "duplex process" for M50 was carried out first with solution heat treatment in air as a manner of decarburization, followed by tempering and plasma nitriding. The details of the duplex processes are described as follows.

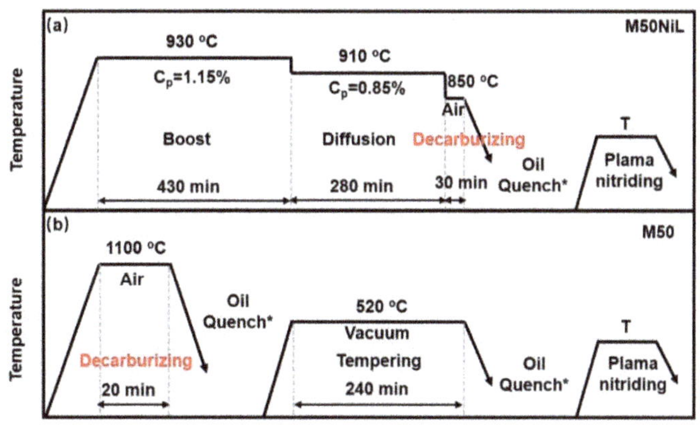

Figure 1. Schematic diagram of duplex process for (**a**) M50NiL steel, (**b**) M50 steel.

(1)   Gas carburizing and decarburizing for M50NiL For M50NiL steel, a "boost-diffusion" gas carburizing and a short decarburizing step were performed at continuous gas carburizing furnace (FAW Group) in Changchun, China, with the process schematically shown in Figure 1a. For comparison, a group of as-carburized samples (denoted as 'C') were kept at the same temperature and time with decarburizing while the atmosphere maintained with the carbon potential of $C_p = 0.85\%$.

(2)   Decarburizing and tempering for M50 To decarburize M50, specimens were solution treated in a box furnace in air at 1100 °C for 20 min. Then the specimens were encapsulated in a vacuum quartz tube and tempered at 520 °C for 4 h. The duplex process is schematically shown in Figure 1b. For comparison, a "Blank" sample was solution treated in vacuum, followed by tempering at the same condition as the decarburized sample. It should be noted that the oxidized layers of air-decarburized samples were grinded off prior to plasma nitriding.

(3)   Plasma nitriding Plasma nitriding was performed in a 30-kW pulse plasma multi-element furnace (LDMC-30AFZ, 30 kW). The atmospheric pressure was 200 Pa, and the voltage was kept at 650 V. The heating rate was 4 °C/min. M50NiL steel samples were nitrided at 520 °C, 540 °C and 580 °C for 4, 8 and 16 h in a gas mixture of N_2 and H_2 with a flow ratio 1:1. M50 steel samples were nitrided at 500 °C, 520 °C and 540 °C for 4, 8 and 12 h in a gas mixture of N_2 and H_2 with a flow ratio 1:3. After nitriding, the specimens were furnace cooled to room temperature in N_2 flow with a cooling rate of ~2 °C/min.

For clarification, detailed processing conditions and corresponding sample denotations for M50NiL and M50 samples are listed in Tables 1 and 2.

Table 1. Duplex process for M50NiL.

Sample Denotation	Carburizing		Decarburizing	Plasma Nitriding
	Boost	**Diffusion**		
C			850 °C, 30 min, C_p = 0.85%	-
CD	930 °C, 430 min, C_p = 1.15%	910 °C, 280 min, C_p = 0.85%	850 °C, 30 min, air	
C-N			850 °C, 30 min, C_p = 0.85%	520 °C, 540 °C, 580 °C 4 h, 8 h, 12 h $N_2:H_2$ = 1:1
CD-N			850 °C, 30 min, air	

Table 2. Duplex process for M50.

Sample Denotation	Solutioning/Decarburizing	Tempering	Plasma Nitriding
Blank	1100 °C, 20 min, vacuum	520 °C, 4 h, vacuum	-
D	1100 °C, 20 min, air		
N	1100 °C, 20 min, vacuum		500 °C, 520 °C, 540 °C 4 h, 8 h, 12 h $N_2:H_2$ = 1:3
D-N	1100 °C, 20 min, air		

2.3. Microstructure Characterizations

The cross-sections of the modified layers were metallographically polished and etched by 4% Nital solution, and then observed by optical microscopy (OM, CMM-33E, Jinan Fengzhi Test Instrument Co., Ltd., Jinan, China). The nitrided layer thickness was identified by the etched boundary evident in OM observations. The phase structures in the surface layer were analyzed by the X-ray diffractometer (type D/max-rB) with Cu-Kα radiation (λ = 0.15406 nm) and by transmission electron microscopy (TEM, JEM-2100, JEOL, Tokyo, Japan) in the bright field and selected area diffraction (SAD) modes. TEM specimens were prepared by single-sided ion thinned from the base side without damage to the nitrided surface. The microhardness on the surface and along the cross-section was measured by Vickers hardness tester (type HV-1000, Shanghai Jvjing Precision Instrument Manufacturing Co., Ltd., Shanghai, China) under a load of 100 g and a dwelling time of 15 s. Three indentations at each depth were taken and averaged. All tests were conducted in air.

3. Results and Discussion

3.1. Microstructures of Modified Surface Layers of M50NiL Steel

Figure 2a is an optical micrograph of the cross-section of Sample C showing that excessive carbon forms carbide (bright) primarily at prior austenite grain boundaries during carburizing. The region enclosed by the carbides has a characteristic scale of around 50 μm, corresponding to the prior austenite grain size. X-ray diffraction of Sample C outmost surface in Figure 2b proves the carbides to be β-Mo_2C and Cr_7C_3, which is consistent with the predicted stable phases of M50NiL at the carburizing temperature (910 °C~920 °C) when the carbon content is greater than 0.85 wt.% (using TCFE9 thermodynamic database). For Sample CD (decarburized in air) in Figure 2c, carbides along the prior austenite grain boundaries coarsen, but not forming into significant networks. The hardness profiles of M50NiL Sample C and Sample CD along the diffusion direction in Figure 2d reflects the bulk diffusion of carbon atoms during carburizing and decarburizing. After carburizing, the carburized case thickness reaches about 1.2 mm. The surface hardness is ~720 $HV_{0.1}$ and the peak hardness (~800 $HV_{0.1}$) offsets to about 350 μm from the outmost surface due to the presence of retained austenite as indicated by X-ray diffraction in Figure 2b. For the decarburized sample, the surface hardness is as low as the matrix and the hardness peak is closer to the surface, manifesting the thickness of decarburized layer is about 100~200 μm where the hardness is lower than the carburized layer.

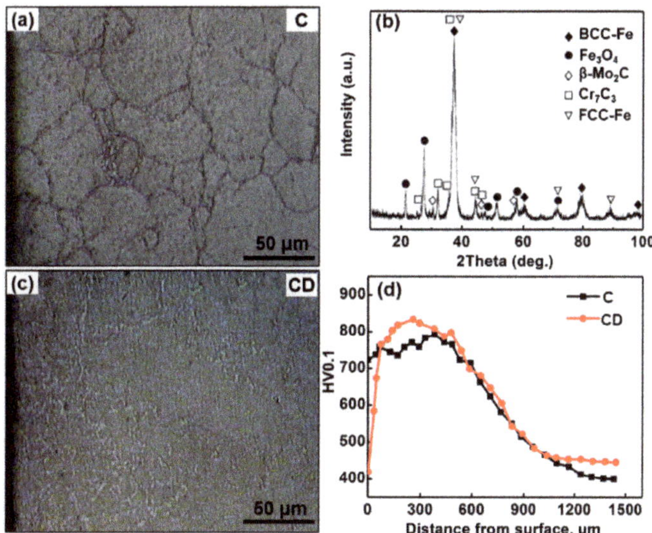

Figure 2. (**a**) Optical micrograph of cross-sectional microstructure of the carburized layer of Sample C; (**b**) XRD pattern of Sample C surface; (**c**) Optical micrograph of cross-sectional microstructure of the carburized layer of Sample CD; (**d**) Microhardness profiles along the modified layers of Sample C and Sample CD.

Figure 3 represents the cross-sectional micrographs and microhardness profiles of the nitrided layers of M50NiL samples nitrided at 520 °C for 4 h. The nitrided layer is etched dark close to the surface. For Sample C-N that is the directly nitrided over the carburized case, the nitrided layer thickness is only 75 μm in Figure 3a, half of the thickness of Sample CD-N (see Figure 3c) that is decarburized prior to plasma nitriding. This indicates that the pre-existing C atoms may remarkably hinder the diffusion of N atoms. The hardness profile of the nitrided case in Figure 3b shows that the surface hardness of Sample CD-N can reach ~1000 HV, whereas the counterpart of Sample C-N is only ~750 HV, slightly higher than the hardness of the carburized case as in Figure 2d. According to Figure 3d that shows the diffraction patterns of the nitrided samples, Sample C-N contains large amount of FCC-Fe phase, which refers to retained austenite in the carburized case. Prior work [15] on gas nitriding of maraging steel with different amount of retained austenite demonstrates that the hardness of the nitrided layer drops with the increased amount of retained austenite. By analyzing the intensity of the peaks in Figure 3d, the percentage of austenite in the surface of Sample CD-N nitrided at 520 °C for 4 h is about 10% while the percentage is 15% in the surface of Sample C-N nitrided at 520 °C for 4 h. This agrees with the findings in [15], and it can further support that the nitrided case hardness and nitrided layer thickness can be enhanced when C atoms and therefore retained austenite are partially removed from the surface layer.

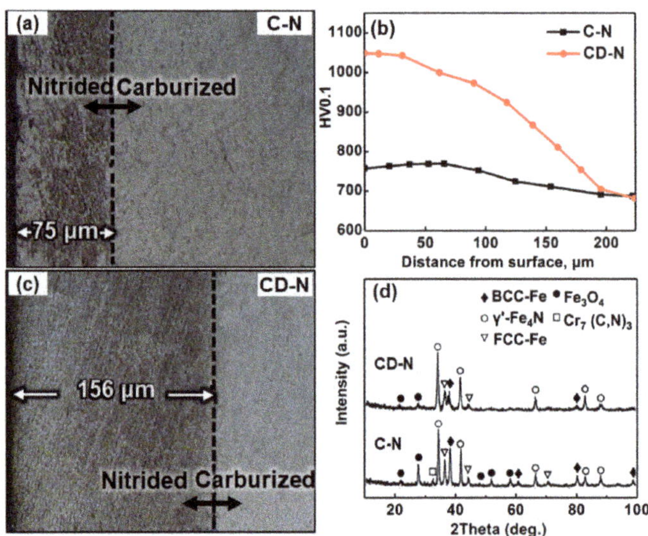

Figure 3. (**a**) Optical micrograph of cross-sectional microstructure of the nitrided layer of M50NiL Sample C-N nitrided at 520 °C for 4 h; (**b**) Microhardness profiles along the modified layers of Sample C-N and Sample CD-N nitride at 520 °C for 4 h; (**c**) Optical micrograph of cross-sectional microstructure of the nitrided layer of M50NiL Sample CD-N nitrided at 520 °C for 4 h; (**d**) XRD patterns of Sample C-N and CD-N nitride at 520 °C for 4 h.

3.2. Microstructures of Modified Surface Layers of M50 Steel

Similar improvement in nitrided layer thickness can be observed in pre-decarburized M50 samples. According to Figure 3a,b, the nitride layer thickness of pre-decarburized M50 Sample D-N nitrided at 500 °C for 4 h is 107 μm, 40% greater than that of Sample N, which was directly nitrided at the same plasma nitriding conditions without decarburization. This indicates that for high-carbon steels, partly removing the solutioned C atoms can promote the nitriding kinetics. Meanwhile, the surface hardness of the nitrided layer for Sample D-N nitride at 500 °C for 4 h reaches 1200 $HV_{0.1}$ which is more significantly improved compared with the non-decarburized sample.

According to XRD patterns in Figure 4c, the peaks of M50 matrix significantly shift to the right after decarburization. This indicates the shrinkage of Fe lattice by the removal of solutioned carbon atoms. By comparing the (110) peak of M50 samples with and without decarburization after tempering as in the inset of Figure 4c, it can be observed that besides shifting, the peak is greatly narrowed after decarburization, which reflects the relaxation of the residual stresses in the surface. The residual compressive stresses induced by martensitic transformation is parallel to the surface, which can hinder nitrogen diffusion perpendicular to the surface upon nitriding [16]. I. Calliari [17] found the white-color microstructure appearing in the nitrided layer is caused by decarburization. As suggested by Figure 4d, after nitriding for 4 h at 500 °C, a low-nitrogen compound $FeN_{0.076}$ is detected in the surface of Sample D-N in contrast to the formation of traditional γ'-Fe_4N in Sample N that is directly nitrided without decarburization. The formation of $FeN_{0.076}$ in the nitrided layer during plasma nitriding is proved to exhibit high hardness [18], which therefore can interpret the high hardness of D-N samples.

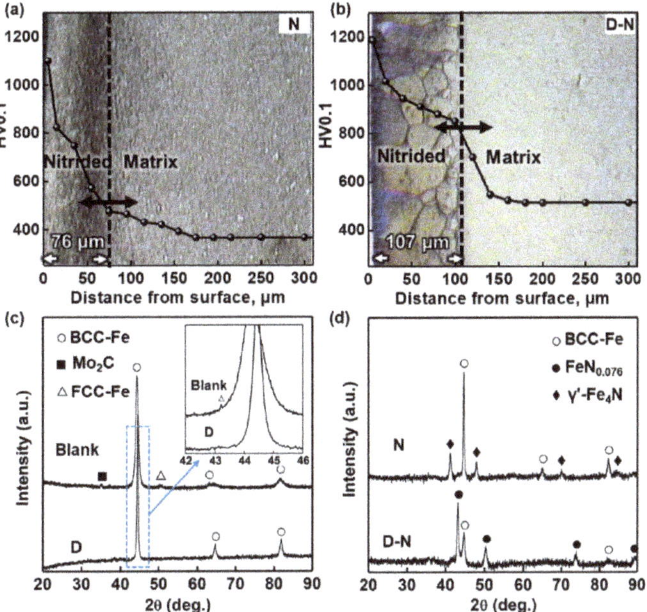

Figure 4. (**a**) Optical micrograph of cross-sectional microstructure of M50 Sample N nitrided at 500 °C for 4 h; (**b**) Optical micrograph of cross-sectional microstructure of M50 Sample D-N nitrided at 500 °C for 4 h; (**c**) XRD patterns of the surface of M50 Sample 'Blank' and Sample D without nitriding; (**d**) XRD patterns of M50 Sample N and Sample D-N nitrided at 500 °C for 4 h.

TEM image and selected area electron diffraction (SAED) pattern of the surface of M50 Sample D-N nitrided at 500 °C for 4 h is shown in Figure 5. Consistent with the phase structure of Sample D-N identified by XRD in Figure 4b, $FeN_{0.076}$ and BCC-Fe are indexed as the primary phases in the nitrided surface. Moreover, SAED pattern is an indicator of evident microstructure refinement through concentric rings, collection of large number of crystals with different orientations. The morphology of refined microstructure is manifested in Figure 5a, which exhibits the light and dark parallel stripes typical spinodal decomposition character [19]. In our previous study, a thermodynamic model has been created to use the limit of thermal stability of solid solution to explain the mechanism of microstructure refinement during nitriding as spinodal decomposition [20]. With spinodal decomposition, conventional γ'-Fe_4N nitride is compressed by the formation of nitrogen-lean $FeN_{0.076}$ and nitrogen-rich martensite. In this way, refined microstructure can enhance the diffusion of nitrogen atoms along highly dense grain boundaries [21].

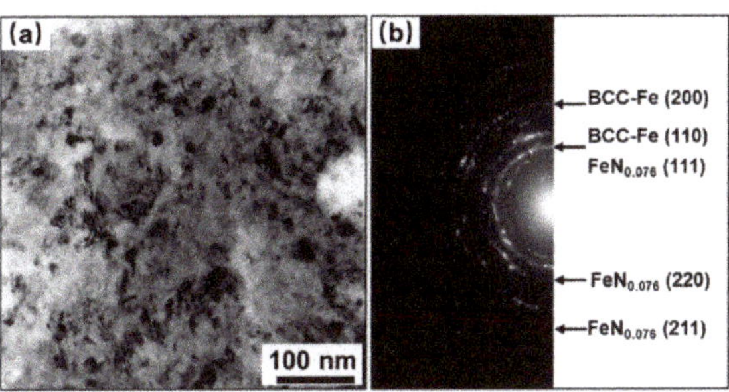

Figure 5. (**a**) TEM image and (**b**) the corresponding SAED of the surface of the M50 Sample D-N nitrided at 500 °C for 4 h.

3.3. The Kinetics of Nitrogen Diffusion during Plasma Nitriding

Since surface can be easily saturated with the nitrogen species during plasma bombarding the surface, and no white nitride layer was observed in M50 or M50NiL nitrided samples, nitrogen diffusion into the matrix is the primary rate-limiting step [16,22]. Therefore, the kinetics of nitriding follows the parabolic rate law, i.e., the thickness of the nitrided layer (Δ) can be expressed by Equation (1) [23],

$$\Delta^2 = k_1 D_N t \tag{1}$$

where D_N (in m^2/s) is the diffusivity of N in steel, t (in s) is the nitriding time, and k_1 is a concentration-dependent constant. The dependence of D_N on temperature is described by the Arrhenius relationship [24,25], i.e.,

$$D_N = D_0 e^{-(\frac{Q}{RT})} \tag{2}$$

where D_0 is the frequency factor (pre-exponential constant), Q (in J·mol^{-1}) is the activation energy, R is the gas constant (8.314 J·mol^{-1}·K^{-1}), and T (in K) is temperature. D_0 and Q can be determined by the linear fitting of ln D_N and $1/T$.

By analyzing the nitride layer thickness of all M50 and M50NiL samples that were nitrided at different nitriding temperatures and time as indicated in Tables 1 and 2, a linear relation between the calculated ln D_N and the reciprocal of nitriding temperatures was obtained for M50 and M50NiL samples, respectively, as shown in Figure 6. The slope of the fitted lines as labeled in Figure 6 reflects the activation energy required for nitrogen atoms to diffuse upon plasma nitriding, and it can imply carbon's hindering effect on nitrogen atoms. It is reasonable to find that M50NiL Samples and M50 N samples that are not pre-decarburized exhibit larger activation energy compared with pre-decarburized samples due to the carbon barrier effect. Moreover, compared with the activation energy of nitrogen atom to diffuse in bcc-lattice (-74.791 kJ/mol) [26], the values obtained in this work is much smaller. This implies that the reduced activation energy is resulted fromincreased number density of grain boundary and defects via microstructure refinement during plasma nitriding.

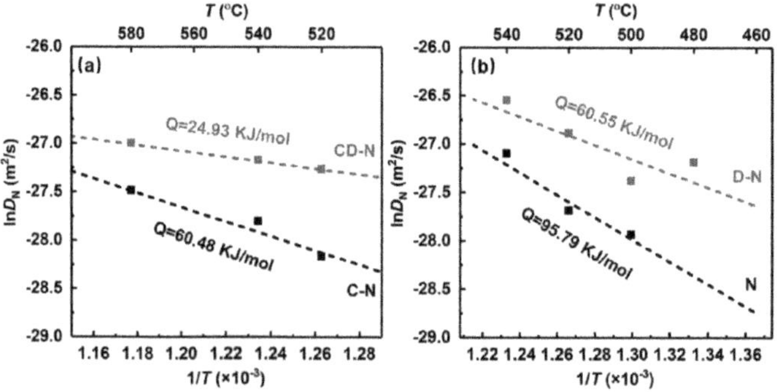

Figure 6. The activation energy for plasma nitrided samples (**a**) M50NiL C-N samples and CD-N samples, (**b**) M50 N samples and D-N samples.

4. Conclusions

In summary, decarburization, which is generally considered detrimental to surface properties, is innovatively applied before plasma nitriding to promote the nitriding kinetics for case-hardened M50NiL and through-hardened M50 bearing steel in this work. The influence of decarburization on properties and growth kinetics of the nitrided layer over the decarburized case is investigated.

(1) Compared with the conventional "duplex surface engine" method, the duplex process with decarburizing step performed in this work can significantly increase the nitrided layer thickness even by more than ~100% and the surface hardness reach ~1200 $HV_{0.1}$.

(2) The analysis of nitriding kinetics shows that the pre-existing carbon atoms solutioned in Fe lattice can hinder the growth of nitride layers. Decarburization performed in air can produce a decarburized case with less carbon and residual compressive stresses, which can significantly increase the thickness of the subsequent nitride layer.

(3) Low-nitrogen compound $FeN_{0.076}$ with high hardness was produced on the modified surface of the decarburizing sample, which indicates that decarburizing can promote the occurrence of surface microstructure refinement via spinodal decomposition during plasma nitriding. Under the same nitriding conditions, the nanostructure can enhance nitrogen diffusion into the matrix along grain boundaries.

Author Contributions: Conceptualization, F.Y.; methodology, F.Y. and J.Y.; validation, F.Y. and J.Y.; formal analysis, F.Y. and J.Y.; resources, M.Y. and Y.X.; data curation, F.Y., J.Y., and B.C.; writing—original draft preparation, F.Y., J.Y., and Y.Y.; writing—review and editing, M.Y., Y.Z., and Y.Y.; visualization, F.Y.; supervision, M.Y.; project administration, M.Y.; funding acquisition, M.Y. and Y.Z. All authors have read and agreed to the published version of the manuscript.

Funding: This research was funded by National Key Research and Development Program of China (2018YFB2001901) and National Key Research and Development Program of China (2017YFB0304601).

Institutional Review Board Statement: Not applicable.

Informed Consent Statement: Not applicable.

Data Availability Statement: Data presented is original.

Acknowledgments: This work was supported by National Key Research and Development Program of China through award numbers 2018YFB2001901 and 2017YFB0304601.

Conflicts of Interest: The authors declare no conflict of interest.

References

1. Bhadeshia, H. Steels for bearings. *Prog. Mater. Sci.* **2012**, *57*, 268–435. [CrossRef]
2. Flodström, I. *Nitrocarburizing and High Temperature Nitriding of Steels in Bearing Applications*; Chalmers University of Technology: Göteborg, Sweden, 2012; pp. 34–36.
3. Ooi, S.; Bhadeshia, H.K.D.H. Duplex Hardening of Steels for Aeroengine Bearings. *ISIJ Int.* **2012**, *52*, 1927–1934. [CrossRef]
4. Rhoads, M.; Johnson, M.; Miedema, K.; Scheetz, J.; Williams, J. Advances in Steel Technologies for Rolling Bearings. Introduction of Nitrided M50 and M50NiL Bearings into Jet Engine Mainshaft Applications. In *Bearing Steel Technologies*; ASTM International: Washington, DC, USA, 2015; Volume 10.
5. Streit, E.; Brock, J.; Poulin, P. Performance Evaluation of "Duplex Hardened" Bearings for Advanced Turbine Engine Applications. *J. ASTM Int.* **2006**, *3*, 1–9. [CrossRef]
6. Bell, T.; Dong, H.; Sun, Y. Realising the potential of duplex surface engineering. *Tribol. Int.* **1998**, *31*, 127–137. [CrossRef]
7. Davies, D. Duplex Hardening: An Advanced Surface Engineering Technique for Helicopter Transmissions. In *Surface Engineering*; Springer: Berlin/Heidelberg, Germany, 1990; pp. 228–237.
8. Tsujikawa, M.; Yamauchi, N.; Ueda, N.; Sone, T.; Hirose, Y. Behavior of carbon in low temperature plasma nitriding layer of austenitic stainless steel. *Surf. Coat. Technol.* **2005**, *193*, 309–313. [CrossRef]
9. Duarte, M.C.; Godoy, C.; Wilson, J.A.-B.; Godoy, G. Analysis of sliding wear tests of plasma processed AISI 316L steel. *Surf. Coat. Technol.* **2014**, *260*, 316–325. [CrossRef]
10. Tsujikawa, M.; Yoshida, D.; Yamauchi, N.; Ueda, N.; Sone, T. Surface Modification of SUS304 Stainless Steel Using Carbon Push-Ahead Effect by Low Temperature Plasma Nitriding. *Mater. Trans.* **2005**, *46*, 863–868. [CrossRef]
11. Tsujikawa, M.; Yoshida, D.; Yamauchi, N.; Ueda, N.; Sone, T.; Tanaka, S. Surface material design of 316 stainless steel by combination of low temperature carburizing and nitriding. *Surf. Coat. Technol.* **2005**, *200*, 507–511. [CrossRef]
12. Bloyce, A.; Sun, Y.; Li, X. Duplex thermochemical processing of M 50 NiL for gear applications. *Heat Treat. Met. Birm.* **1999**, *26*, 37–40.
13. Egert, P.; Maliska, A.; Silva, H.; Speller, C. Decarburization during plasma nitriding. *Surf. Coat. Technol.* **1999**, *122*, 33–38. [CrossRef]
14. Zhao, X.; Guo, J.; Wang, H.; Wen, Z.; Liu, Q.; Zhao, G.; Wang, W. Effects of decarburization on the wear resistance and damage mechanisms of rail steels subject to contact fatigue. *Wear* **2016**, *364*, 130–143. [CrossRef]

15. Hussain, K.; Tauqir, A.; Haq, A.U.; Khan, A.Q. Effect of retained austenite on gas nitriding of high strength steel. *Mater. Sci. Technol.* **1998**, *14*, 1218–1220. [CrossRef]

16. Akbari, A.; Mohammadzadeh, R.; Templier, C.; Riviere, J. Effect of the initial microstructure on the plasma nitriding behavior of AISI M2 high speed steel. *Surf. Coat. Technol.* **2010**, *204*, 4114–4120. [CrossRef]

17. Calliari, I.; Dabalà, M.; Ramous, E.; Zanesco, M.; Gianotti, E. Microstructure of a Nitrided Steel Previously Decarburized. *J. Mater. Eng. Perform.* **2006**, *15*, 693–698. [CrossRef]

18. Yan, M.; Chen, B.; Li, B. Microstructure and mechanical properties from an attractive combination of plasma nitriding and secondary hardening of M50 steel. *Appl. Surf. Sci.* **2018**, *455*, 1–7. [CrossRef]

19. Wang, J.; Zou, H.; Li, C.; Qiu, S.; Shen, B. The spinodal decomposition in 17-4PH stainless steel subjected to long-term aging at 350 °C. *Mater. Charact.* **2008**, *59*, 587–591. [CrossRef]

20. Yao, J.; Yan, F.; Yan, M.; Zhang, Y.; Huang, D.; Xu, Y. The mechanism of surface nanocrystallization during plasma nitriding. *Appl. Surf. Sci.* **2019**, *488*, 462–467. [CrossRef]

21. Sasidhar, K.N.; Meka, S.R. Thermodynamic reasoning for colossal N supersaturation in austenitic and ferritic stainless steels during low-temperature nitridation. *Sci. Rep.* **2019**, *9*, 7996. [CrossRef]

22. Sun, Y.; Bell, T. A numerical model of plasma nitriding of low alloy steels. *Mater. Sci. Eng. A* **1997**, *224*, 33–47. [CrossRef]

23. Zagonel, L.F.; Figueroa, C.A.; Droppa, R.; Alvarez, F. Influence of the process temperature on the steel microstructure and hardening in pulsed plasma nitriding. *Surf. Coat. Technol.* **2006**, *201*, 452–457. [CrossRef]

24. Scheuer, C.; Cardoso, R.; Mafra, M.; Brunatto, S. AISI 420 martensitic stainless steel low-temperature plasma assisted carburizing kinetics. *Surf. Coat. Technol.* **2013**, *214*, 30–37. [CrossRef]

25. Yang, Y.; Yan, M.; Zhang, S.; Guo, J.; Jiang, S.; Li, D. Diffusion behavior of carbon and its hardening effect on plasma carburized M50NiL steel: Influences of treatment temperature and duration. *Surf. Coat. Technol.* **2018**, *333*, 96–103. [CrossRef]

26. Pinedo, C.E.; Monteiro, W.A. On the kinetics of plasma nitriding a martensitic stainless steel type AISI 420. *Surf. Coat. Technol.* **2004**, *179*, 119–123. [CrossRef]

Article

Effect of Deformation on the Microstructure of Cold-Rolled TA2 Alloy after Low-Temperature Nitriding

Guotan Liu [1], Huanzheng Sun [2], Enhong Wang [1], Keqiang Sun [1], Xiaoshuo Zhu [3] and Yudong Fu [1,*]

[1] School of Materials Science and Chemical Engineering, Harbin Engineering University, Harbin 150001, China; liuguotan097@163.com (G.L.); WangEH_2021@163.com (E.W.); skq000163@163.com (K.S.)
[2] School of Materials Science and Technology, Chongqing University, Chongqing 400044, China; 20190901273@cqu.edu.cn
[3] School of Mechanical Engineering, Xinjiang University, Urumqi 830045, China; Zhu-xiaoshuo@hrbeu.edu.cn
* Correspondence: fuyudong@hrbeu.edu.cn

Abstract: In order to improve the low hardness and poor wear resistance of TA2, this paper proposes a composite process of cold-rolling and low-temperature plasma nitriding with recrystallization. This composite modification process can effectively achieve the dual goals of modifying the matrix structure and surface of TA2 alloy simultaneously. The cold-rolling experimental results indicate that when the deformation rate increases, the small-sized grains in the sample increase significantly, and the grain orientation changes from TD (transverse direction) to RD (rolling direction) and then to TD. The nitriding experimental results indicate that the {0001} basal surface texture deflected from the TD direction to the RD direction, {10-10} cylindrical texture components gradually increased, and the special orientation phenomenon of cylindrical and conical texture disappeared, it can be seen that an increase in the deformation rate promotes recrystallization. The XRD test results indicate that after low-temperature nitriding, metastable nitriding phase TiN0.26 is formed on the surface of TA2. The SEM morphology of the cross-section shows that a relatively special nitrided zone is formed, and mechanical performance test results indicate the wear resistance and hardness of the alloy increased.

Keywords: TA2 alloy; cold-rolling deformation; recrystallization; low-temperature nitriding; microstructure; mechanical properties

Citation: Liu, G.; Sun, H.; Wang, E.; Sun, K.; Zhu, X.; Fu, Y. Effect of Deformation on the Microstructure of Cold-Rolled TA2 Alloy after Low-Temperature Nitriding. *Coatings* **2021**, *11*, 1011. https://doi.org/10.3390/coatings11081011

Academic Editor: Michał Kulka

Received: 7 July 2021
Accepted: 19 August 2021
Published: 23 August 2021

Publisher's Note: MDPI stays neutral with regard to jurisdictional claims in published maps and institutional affiliations.

Copyright: © 2021 by the authors. Licensee MDPI, Basel, Switzerland. This article is an open access article distributed under the terms and conditions of the Creative Commons Attribution (CC BY) license (https://creativecommons.org/licenses/by/4.0/).

1. Introduction

Titanium and its alloys have attracted great attention in the fields of scientific research and industrial production due to their outstanding advantages, such as excellent heat resistance, corrosion resistance, biocompatibility, and high specific strength [1–3]. However, the low surface hardness and poor wear resistance of titanium alloys greatly limit its application. The TA2 titanium alloy in this paper belongs to α-type titanium alloy [4–6], and this type of alloy cannot be strengthened by a heat treatment process, so in practice, the strength is average in the annealed state [4,7].

Rolling deformation is a commonly used forming technology for TA2 industrial titanium alloy sheets [8,9]. Gurao et al. [10] studied the evolution characteristics of the texture and microstructure of industrial pure titanium alloys under different rolling paths through experiments and self-consistent simulations. S.K. Sahoo et al. [11] studied the texture and microstructure evolution of high-deformation hot-rolled industrial pure titanium under a specific rolling method. The experimental results indicate that the basal surface texture in the alloy sample does not change with the rolling type and the deformation reduction.

After deformation, the hardness increases and plastic toughness decrease for the metal material, which is not conducive to subsequent mechanical processing [11,12]. To facilitate processing, it is usually to anneal and recrystallize the deformed metal to soften the material. With the increase of time and temperature, the annealing recrystallization process is often divided into three stages: recovery, recrystallization, and grain growth [13].

Researchers have continued to deepen the research on densely packed hexagonal metals [14–17]. However, there are a few studies on the recrystallization of hexagonal metals, such as titanium. R. J. Contieri et al. [18] studied the recrystallization behavior of cold-rolled titanium plates. Experimental data indicate that the dislocation density of titanium plates increases significantly after cold rolling. When the recrystallization annealing temperature is 650–670 °C, the average recrystallization activation energy is about 156.8 KJ/mol. Hayama et al. [19,20] studied the annealing behavior of cold-rolled titanium alloys and found that 50% and 70% deformation rate specimens can be annealed at 700 and 800 °C for 15 min to achieve an equiaxed crystalline state.

In recent years, the modification of the mechanical properties of titanium alloys has been a research hotspot in the field of materials. Nitriding is one of the important surface modification technologies for metal materials. Nitriding gives metal materials higher surface hardness and strength without shape and size changes. However, in engineering applications, the nitriding temperature of titanium alloys is usually between 800 and 900 °C. At this temperature, coarse grains, severe high-temperature oxidation, and deep hydrogen diffusion occur. In order to avoid these problems, a lower nitriding temperature is a hopeful solution; therefore, researchers have investigated promoting penetration. H. Kovacl et al. [21,22] studied shot peening as a nitriding promotion process for AISI4140 steel by the comparison experiments of shot peening with different densities. The experimental results indicate that shot peening forms finer crystal grains on the surface of the sample, and the increase of the surface dislocation density accelerates the diffusion of nitrogen atoms. The thickness of the infiltration layer of the sample treated by the shot peening and plasma composite process is almost twice the nitriding treatment alone, and the surface hardness also significantly increases. Liu Rui-Liang et al. [7,23] studied the rare earth infiltration process and found that the surface phase structure of the 17-4PH martensitic stainless steel after the rare-earth carbonitriding does not change significantly, mainly due to the expansion of martensite containing C and N Body α'N, γ'-Fe4N, and CrN are equal in composition. However, after adding rare earth, the structure of the infiltrated layer is denser, and the thickness of the infiltrated layer increases by nearly 50%, also the hardness of the infiltrated layer is increased by about 100 HV and the wear resistance is also significantly improved. In M50NiL steel, the surface hardness increases by 143 HV0.1 because of the addition of rare-earth atoms and the thickness of the nitrocarburized layer increases by 39 μm. As a pretreatment method to accelerate the nitriding process, cold-rolling deformation can theoretically not only refine the substructure of the surface layer of the metal material but also increase the dislocation density inside the metal material and increase the storage distortion energy, thereby reducing the N diffusion activation energy of the atom infiltration. At present, the research on deformation promotion technology is still in the experimental stage, and the materials are mostly steel. However, the research papers on the deformation promotion of hexagonal close-packed metals are rarely reported.

Based on the research in the past few years [24–26], we figured that low-temperature ($\leq$550 °C) plasma-nitriding treatment could be used as the preferred surface modification process for titanium alloy, and cold-rolling deformation can promote plasma nitriding. In this paper, a treatment process of cold-rolling deformation and low-temperature nitriding at 450 °C, causing simultaneous recrystallizations, was designed to perform compound modification treatment on TA2 industrial pure titanium alloy. Under cold deformation conditions, a large number of defects (such as point defects and dislocations) are generated, and the lattice distortion energy stored in the polycrystalline metal continuously increases; these promote the diffusion and migration of N atoms during the low-temperature ion nitriding process. The temperature of the nitriding was designed within the recrystallization temperature of TA2 to achieve the effect of low-temperature nitriding composite recrystallization. This process can not only realize the control of the alloy matrix structure but also effectively improve the wear resistance of the alloy. In this paper, the evolution of the microstructure after cold rolling, the impact of deformations on the recrystallization, the evolution of the recrystallized matrix under different deformations, the surface phase structure, and wear resistance after low-temperature nitriding are investigated.

2. Materials and Methods

The TA2 alloy used in this paper belongs to commercial titanium, and its chemical composition (wt%) is 0.106% Fe, 0.0155% C, 0.0135% N, 0.0005% H, 0.1165% O, and the balance is Ti. The metallographic structure is composed of fully annealed recrystallized phases. Plastic deformation occurs at room temperature, and the sample size is $15 \times 15 \times X$ mm. The length of the sample is parallel to the rolling direction (RD), and the sample height X is 20 mm (0% deformation rate), 16 mm (20%), 12 mm (40%), 8 mm (60%), and 4 mm (80%), respectively. The incoming TA2 samples with a thickness of 20 mm were rolled several times at room temperature, but the maximum reduction did not exceed 0.2 mm/time. After cold rolling, the plasma nitriding was performed at 450 °C for 16 h in a pulsed plasma furnace with a voltage setting of 650 V, and the atmosphere's ratio of H2 and N2 is 5:7.

The experimental samples were mechanically polished in order with 400, 800, 1500, and 2000 sandpaper and ultrasonically cleaned and then placed in the furnace. The nitrided samples were polished and ultrasonically cleaned. The samples used for OM\SEM\EBSD have been chemically corroded (corrosive solution $HF:HNO_3:H_2O = 2:1:17$) after mechanically polishing. The specimens for EBSD characterization were electro-polished in an electrolyte consisting of 6 mL perchloric acid, 34 mL n-butanol, and 60 mL methanol with a direct current of 1 A for 1 min. An OLYMPUS-PM3 (OLYMPUS, Fukuoka, Japan) metallographic microscope was used to observe the microstructure of the initial and cold-rolled samples. A JEOL JSM-6480A scanning electron microscope (SEM, Thermo Fisher Scientific, JEOL, Tokyo, Japan) equipped with an energy spectrum analyzer (EDS) was used to analyze the micro-area composition. The acceleration voltage of the EDS test was 20 kV. An X-Pert Pro XRD analyzer (Rigaku, Japan) was used to study the phase composition of the nitride layer, and a copper target continuous scan mode was used, where the scan rate was 5°/min and the range of 2θ was 20 to 80°. The original microstructure, cold-rolled microstructure, and recrystallized microstructure were measured using TESCAN MAIA3 ultra-high-resolution field emission scanning electron microscope (EBSD) (TESCAN, Brno, Czech Republic), the acceleration voltage of EBSD test was 20 kV, scanning step sizes of 1.6 and 0.6 µm were adopted for deformed and nitrided samples, respectively. The test data is analyzed with Channel 5 software. By analyzing the data collected in the experiment, the crystal grain size, shape characteristics, grain boundary distribution, and orientation of the crystal can be obtained. The indentation hardness is measured by the HVS-1000 hardness tester (Times Mountain, Beijing, China) with the load setting of 50 g and a dwell time of 15 s. Five measurements were performed for each test sample, and the values were read 10 times. The abrasion test was carried out by a Pin-On-Disk-1-Auto testing machine (Zhongke Kaihua, Lanzhou, China) with GCr15 grinding ball material. It was set under a load of 5 N, a wear rate of 200 r/min, and a wear time of 20 min.

3. Results and Discussion

3.1. Initial Microstructure of TA2 Alloy

As shown in Figure 1, the initial microstructure of the TA2 alloy sheet before cold deformation is listed with the magnifications of 200 and 500 times, respectively. The initial material is a fully annealed recrystallized structure, without deformed or twin structures, and the grain size is uniform with an average value of about 22 µm.

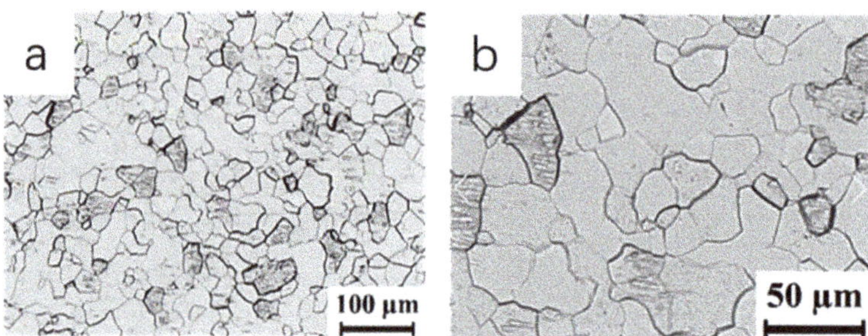

Figure 1. Initial microstructure of TA2 alloy sheet before cold deformation. (**a**) 200× (**b**) 500×.

In Figure 2, the surface shown is the rolling surface, and the horizontal direction is the rolling direction, as well as the vertical direction, is the normal direction of the rolling plane. Figure 2a is the contrast map of the initial state TA2 alloy sheet. The initial structure in Figure 2a belongs to the typical annealed equiaxed structure, and there is no lamellar twin structure. This is consistent with the observation of the microstructure. In Figure 2b, {0001}‖ND crystal grains account for a relatively high proportion; the transition color crystal grains account for the largest proportion, indicating that the c-axis of most crystal grains deviates from the ND direction.

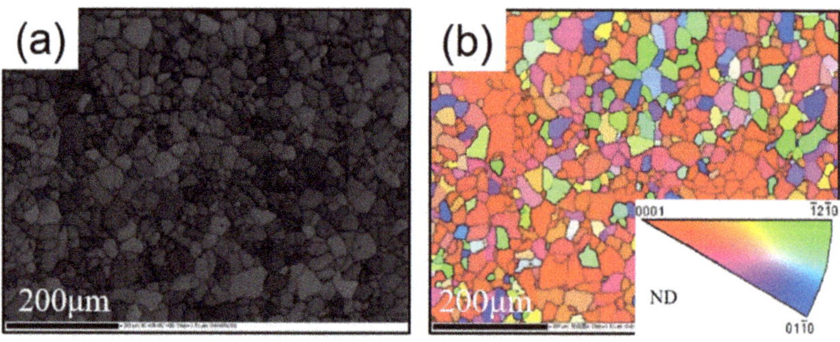

Figure 2. EBSD data of incoming sheet material. (**a**) contrast map and (**b**) grain orientation map.

3.2. Microstructure of Cold-Rolled TA2

The microstructure of TA2 alloy sheets with different cold-rolling deformation (20%, 40%, 60%, and 80%) is shown in Figure 3, and the magnifications are 200 times and 500 times, respectively. Through these, the evolution of the alloy microstructure caused by the degree of deformation is investigated. Compared with the undeformed microstructure in Figure 1, the increase in the deformation rate makes the grains elongated in the rolling direction, and the unevenness of the microstructure is gradually increased. For the 20% deformed sample, as shown in Figure 3a,b, due to its low degree of deformation, there is not much change compared with the initial sample; only a small part of the deformed structure exists, and the grain distribution is relatively even. Among them, the 20% and 40% deformed samples have a smaller degree of grain boundary damage, and the percentage of complete grains is higher than that of the bigger deformed samples. When the deformation rate increases to 40%, the deformation degree of the grains increases significantly. When the deformation exceeds 40%, as shown in Figure 3e–h, the grain boundaries are severely broken, and almost no complete grains are observed, so the degree of grain fragmentation and refinement increases significantly, when deformation reaches the rolling limit state, as shown in Figure 3g,h, a typical cold-rolled deformation microstructure forms.

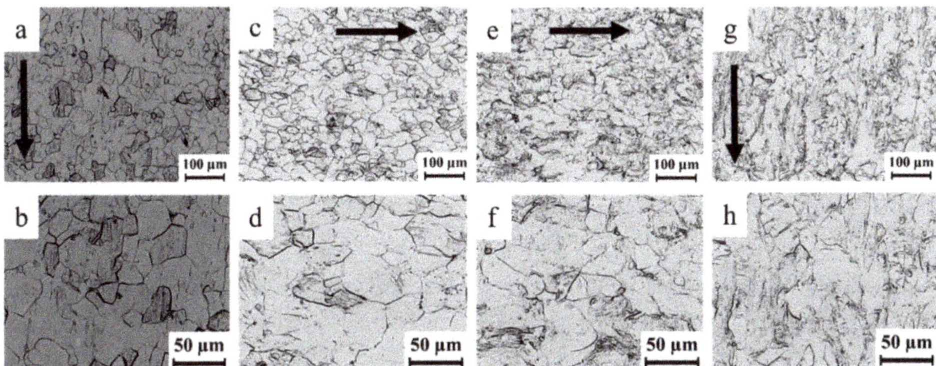

Figure 3. Metallographic structure of TA2 alloy with different deformation: (**a,b**) 20%, (**c,d**) 40%, (**e,f**) 60%, (**g,h**) 80%.

The EBSD pole figures of the 0%, 40%, and 80% deformed samples are shown in Figure 4. Basically, the cold-rolled deformed titanium alloy still maintains a typical bimodal distribution. When the deformation reaches 40%, the basal surface texture deflects by 90° from the TD to the RD direction, resulting in a significant orientation change. In addition, almost no basal texture component forms in the ND direction. When the deformation rate increases to 80%, the texture orientation changes 90° from RD to TD, and the concentration of texture distribution is significant.

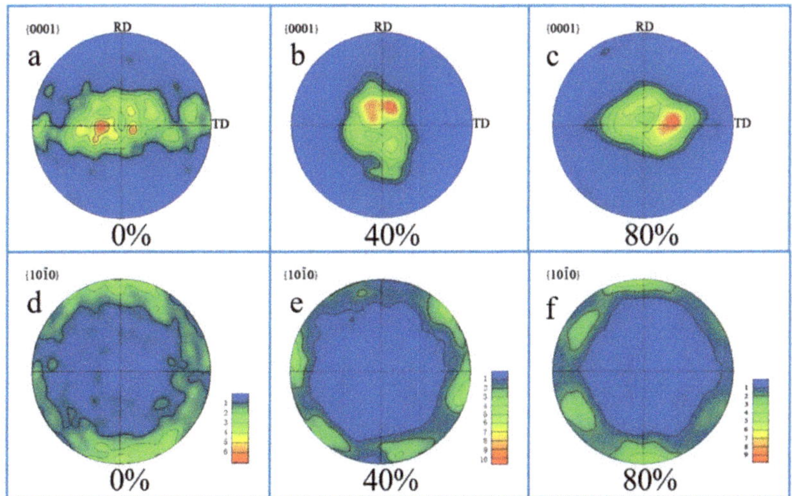

Figure 4. Pole figure of cold-rolled TA2 with different degrees of cold-rolled deformation: (**a,d**) 0% deformed, (**b,e**) 40% deformed, (**c,f**) 80% deformed.

Compared with the 0% deformed samples, the strength of the cold-deformed texture structure significantly increases, and the texture strengths of the 40% and 80% deformed samples are 10 and 9.9, respectively. In the {10-10} pole figure, the intensity of the texture component in the {10-10}‖RD direction presents an increasing trend with the increase of the deformation rate. Compared with the undeformed samples, the degree of "stray" of the cylindrical texture and cone texture distribution of the 40% and 80% deformed samples significantly reduces, indicating a preferred orientation.

3.3. Characterization of Microstructure after Deformation and Nitriding

Conventional nitriding will cause the crystal grains of the titanium alloy to coarsen and cannot form a good matrix structure [27,28]. Next, we studied the microstructure changes of the matrix after cold rolling and low-temperature nitriding. Figure 5 shows the grain orientation map of TA2 alloy with different degrees of cold rolling. Figure 6 shows the grain orientation map and reverse pole map of 40% and 80% deformed samples after nitriding treatment (450 °C × 16 h). In Figure 6a, the proportion of grains in the {0001} direction (red color) significantly reduces compared to Figure 5a, while the number of green and blue grains slightly increases. In addition, the number of transition color grains (as shown by the blue arrow in Figure 5a) increases significantly. This indicates that grain orientation rotates from the {0001} direction to the {01-10} direction, which is an obvious recrystallization state but with a low crystallization nucleation rate. In addition to the change in the orientation of grains, the elongation of the grains in the structure is significant, which is consistent with the changes in the grain size in the upper section. Figure 6b presents the grain orientation map and the inverse pole map of the 80% deformed sample after nitriding. The number of complete fine grains in Figure 6b significantly increases compared to Figures 5b and 6a, and the transition color grains significantly reduce, indicating the alloy structure has a higher degree of recrystallization, and the structure in the blue box is completely equiaxed. In addition, the number of grains in the {0001} direction significantly reduces, and the deformed structure and substructures along the arrow direction still exist but are significantly reduced compared to the 40% nitriding sample.

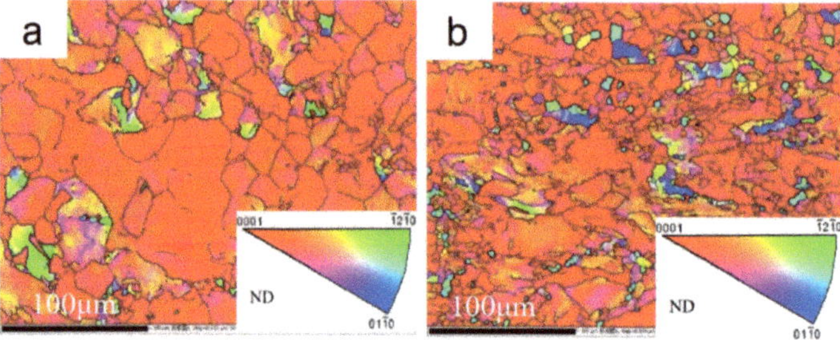

Figure 5. Grain orientation map of TA2 alloy with different degrees of cold-rolling deformation. (**a**) 40% and (**b**) 80%.

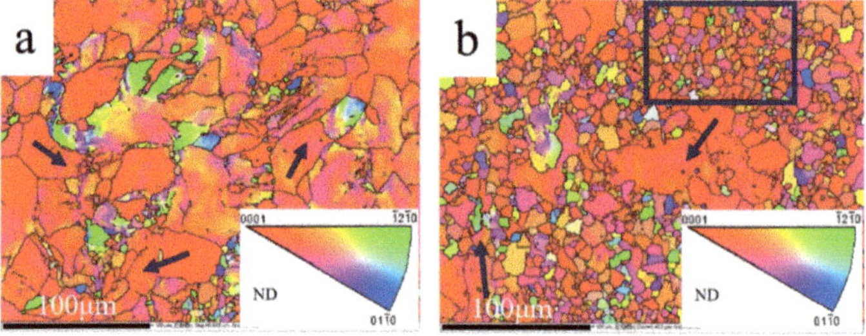

Figure 6. Grain orientation map of cold-rolled deformed TA2 alloy sample after nitriding treatment. (**a**) 40% and (**b**) 80%.

3.4. The Influence of Nitriding on Texture

Figure 7 shows the pole figures of cold-rolled deformed TA2 alloy after nitriding. Compared with Figure 4b,c,e,f, it can be seen that after the 40% deformed sample undergoes nitriding treatment, the {0001} texture parallel to the RD direction has undergone a significant orientation change, from being parallel to the RD direction to being parallel to the TD direction, and the position of the maximum pole density is also approximately rotated by 90°. Only a few crystal grains on the TD-ND plane have a c-axis perpendicular to the RD direction, and most of the crystal grains on this plane have a c-axis deviated from the ND direction by 45–90°, {10-10} the degree of dispersion of the cylindrical texture in the RD-TD plane increases, and the texture changes along the ND direction are relatively subtle; the basal texture {0001} of the 80% deformed sample after nitriding treatment showed double peaks in the RD-TD plane, but the highest strength of the texture was reduced from 9.87 to 8.55 compared with the cold-rolled state. In addition, the divergence of the cylindrical texture {10-10} has increased significantly, and the texture of {10-10}∥RD that originally existed at the extreme position of the RD direction is almost nonexistent, and the texture of the special angle orientation is greatly reduced, and the new texture appears at the pole position in the TD direction. Compared with the 40% deformed sample after nitriding, the {10-10} cylindrical texture component in the ND direction gradually increases.

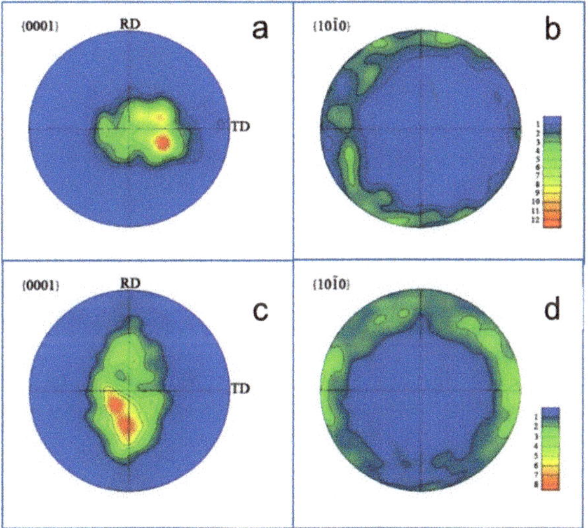

Figure 7. EBSD pole figures of cold-rolled deformed TA2 alloy after nitriding. (**a,b**) 40% deformed and (**c,d**) 80% deformed.

3.5. The Phase Composition after Low-Temperature Nitriding

Figure 8 shows the surface XRD pattern of different deformation rate samples after nitriding at 450 °C for 16 h. In Figure 8 that the main phases contained in the surface layer of the TA2 sample after nitriding for 16 h are α-Ti and TiN0.26. This is similar to F.S. Braz's research results [29], who also obtained the same compound in the 450 °C nitridation experiment. Of course, the surface may actually contain steady-state nitride, but it was not detected because of its small content. After nitriding, the metastable titanium nitride compound phase TiN0.26 forms on the surface of the 40%, 60%, and 80% deformed samples. The TiN0.26 phase peak of the 20% deformed sample is almost as weak as the 0% deformed sample. The reason is that the small deformation does not produce enough defects inside the sample, resulting in a poor nitriding effect. Different samples of other deformation rates, the α-Ti's peak positions of the 40% deformed sample shift to near 35 and 63°. In the process of low-temperature nitriding of cold-rolled alloy samples, the orientation of the {0001} basal plane texture changes, and the c-axis direction of the grains shifts. According to the Bragg equation (2dsinθ = nλ), the change of interplanar spacing (d) will affect the change of "θ" angle; these are the reasons for the peak position shift.

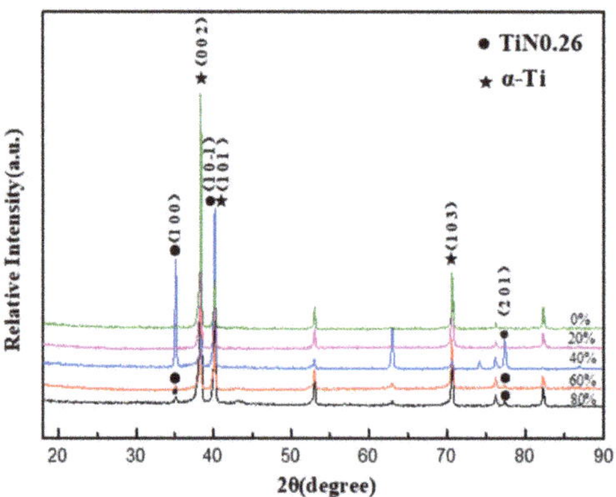

Figure 8. Surface XRD pattern of series of deformed alloy samples after nitriding at 450 °C for 16 h.

In Figure 9, the EDS point analysis of the chemical composition of the 0%, 20%, 40%, 60%, and 80% deformed samples after 450 °C for 16 h nitriding treatment are listed. The point scanning positions are marked in each sub-picture, where spherical particulate matter exists. The element content in the scanning area is shown in Table 1. In Table 1, it indicates that with the increase of the deformation rate of TA2 samples, the mass percentage of N element first increases and then decreases, and the nitrogen-titanium atomic ratio also indicates, at first, an increasing and then decreasing trend. After the deformation rate exceeds 40%, the nitrogen atom content of the white spherical particulate matter decreases because the increase in defect density and the weakening of the interaction force between atoms accelerates the diffusion of active nitrogen atoms, and more N on the surface diffuses into the inside. The 60% sample has the highest surface nitrogen atom content of 76.07%, and the nitrogen-titanium atom ratio is about 3:1. In addition, the bigger the size of the white particles selected at special points is, the higher the N element content is.

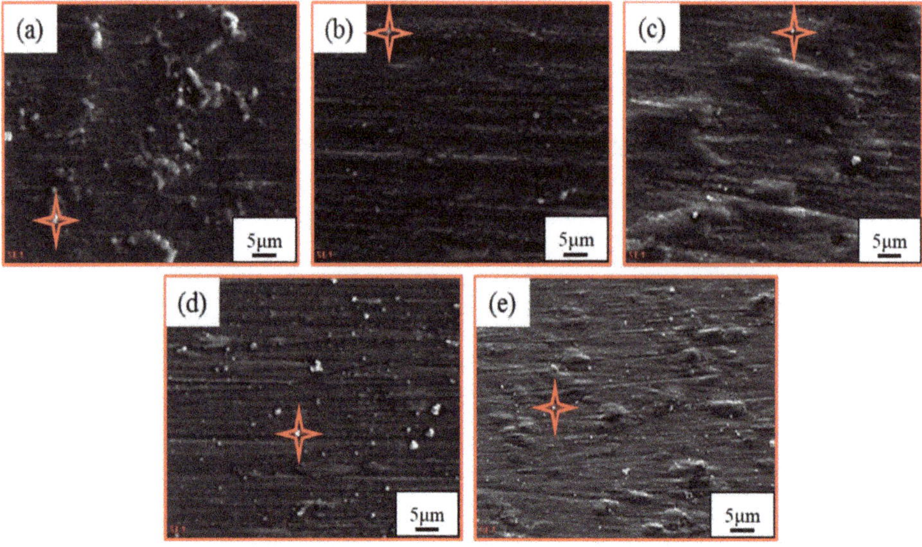

Figure 9. Location selection map of EDS spectrum analysis after TA2 low-temperature nitriding. (**a**) 0% deformed, (**b**) 20% deformed, (**c**) 40% deformed, (**d**) 60% deformed, and (**e**) 80% deformed.

Table 1. EDS Energy Spectrum Test Data of Special Points After Nitriding Series of Deformed Alloys.

Deformation	Element (N)		Element (Ti)	
	W (t)%	A (t)%	W (t)%	A (t)%
0%	25.8	54.3	74.3	45.8
20%	30.0	59.5	70.0	40.5
40%	29.3	58.6	70.7	41.4
60%	48.2	76.0	51.8	23.9
80%	23.8	51.6	76.3	48.4

3.6. Morphology of TA2 Alloy Samples under Different Degrees of Cold Deformation

In the traditional high-temperature nitriding process, a brighter color compound layer will be observed in the cross-sectional morphology of the nitriding sample, and continuous nitrides will be distributed near the surface [30]. This article describes a new nitriding process for the recrystallization of TA2 alloy low-temperature nitriding composite matrix. Under the influence of low-temperature conditions and pre-deformation, a nitriding microstructure different from the traditional nitriding process may appear. Figure 10 shows the SEM morphology of the cross-section of TA2 alloy samples after low-temperature nitriding; it can be seen from the figure that there is no bright white nitride layer under various deformations. Combining the XRD and EDS data in Section 3.5, it can be inferred that a single metastable nitride phase should only be formed near the surface. In addition, we speculate that a nitriding zone with a gradient distribution of N elements is formed in the area below the surface because, at lower temperatures, active N atoms can diffuse into the depth of the matrix through defect channels generated by pre-deformation. Finally, the nitriding microstructure generated by the nitriding process described in this article should be a gradient nitriding region, and there is no obvious nitriding layer, but the N element is distributed in a gradient from the surface to the core.

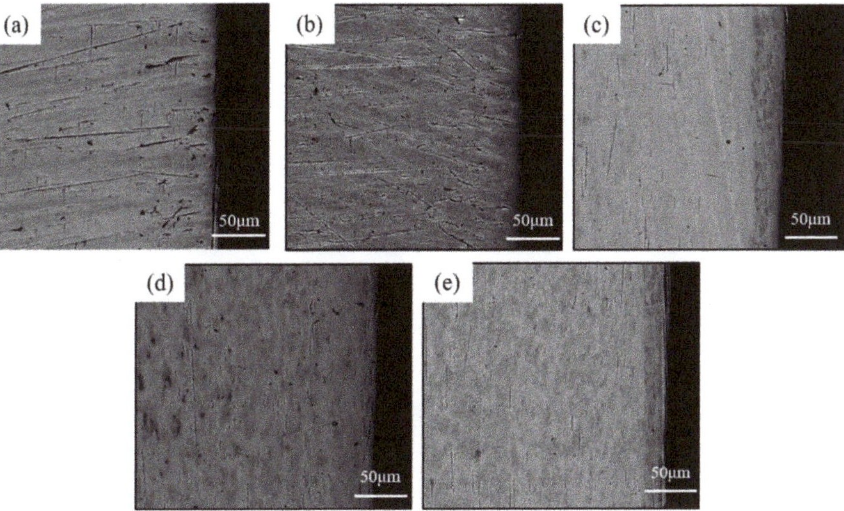

Figure 10. SEM morphology of a cross-section of a TA2 alloy sample after low-temperature nitriding. (**a**) 0% deformed, (**b**) 20% deformed, (**c**) 40% deformed, (**d**) 60% deformed, and (**e**) 80% deformed.

3.7. Research on Mechanical Properties after Low-Temperature Nitriding

Figure 11 shows the friction coefficient curves of alloy samples with different deformation degrees after cold rolling and after nitriding. As Figure 11a shows, after cold rolling, the friction coefficient value of the undeformed sample is stable at about 0.71, which is the maximum value of all the test samples. The minimum coefficient of friction of the deformed sample is 0.48, which is 31.5% lower than that of the undeformed sample. Among them, the friction values of the 40% and 60% deformed samples are stable at about 0.5.

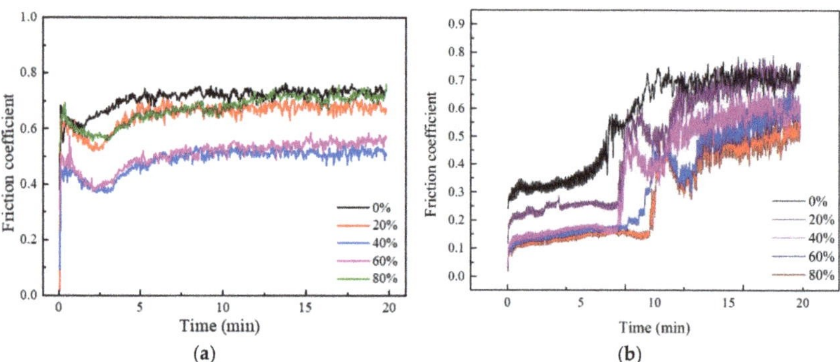

Figure 11. Friction coefficient curves of alloy samples with different deformation degrees after nitriding. (**a**) After cold rolling and (**b**) after nitriding.

As Figure 11b shows, after the nitriding process, the stable friction coefficient indicates a decreasing trend with the increase of the deformation rate. The friction coefficient curve changes at 0%, 20%, and 40% deformation are similar. The stable value of the friction coefficient of the 20% deformed sample has a small difference compared with the 0% deformed sample, which is stable at about 0.7; the stable friction coefficient of the 40% deformed sample is about 0.53. The stable friction coefficient of the 60% deformed sample and the 40% deformed sample are almost the same; the stable friction coefficient of the 80% deformed sample is about 0.5. Both 20% and 40% deformed samples have obvious wear stages with low friction coefficients after nitriding treatment, which are stable at about 0.25 and 0.15, respectively, the duration of this process is about 7 min, while the 0% deformed samples show a steady state at the same time. There is an upward trend, and there is no declining trend after 7 min. When the deformation rate exceeds 40%, the low friction coefficient wear stage takes longer, and the time required for 80% deformed samples reaches 10 min, while the friction coefficient values of the 40%, 60%, and 80% of deformed samples at this stage have little difference, stable at about 0.14. In addition, the friction coefficient of the 20% deformed sample after 14 min of the friction test is the same as that of the 0% deformed sample, which is stable at about 0.7; the friction coefficient of the 40% deformed sample at the same stage is about 0.57, which is 18.6% lower than the 0% deformed sample. The friction coefficient values of the 60% and 80% deformed samples still show an increasing trend after 14 min.

The low friction coefficient wear stage fully proves that the infiltration layer formed on the surface of the cold-rolled deformed alloy sample after the low-temperature nitriding treatment can effectively improve the wear resistance of the alloy, and the increase in the amount of cold-rolled deformation promotes the improvement of the wear resistance. Combining the analysis in the previous section, it can be seen that the reason for the short duration of the low friction coefficient wear stage should be that only a single metastable nitride is formed near the surface, and no continuous nitride region is formed below the surface.

Figure 12 shows the microhardness values of cold-rolled deformed alloy samples before and after nitriding. As Figure 11 shows, after 16 h of nitriding treatment, the microhardness value of TA2 with deformation after low-temperature nitriding treatment is higher than that of the only deformed, of which the 80% deformed samples' increase is the smallest at only 1.7%; the increase of microhardness of other deformation rate is around 8.3%. Comparing the hardness of different deformation rates after low-temperature plasma nitriding treatment, the hardness increases with the increase of the deformation rate. The hardness of the 80% deformed sample has a small increase. The reason is that the nitriding process is synchronized with the recrystallization treatment and the 80% deformed structure is more recrystallized, and the plasticity and toughness of the structure are significantly improved, which in turn causes the microhardness to decrease, this partly negates the effect of nitriding. Combined with the XRD results, the reason for the less increase in surface hardness after nitriding is also due to the formation of metastable nitrides with a lower hardness on the surface.

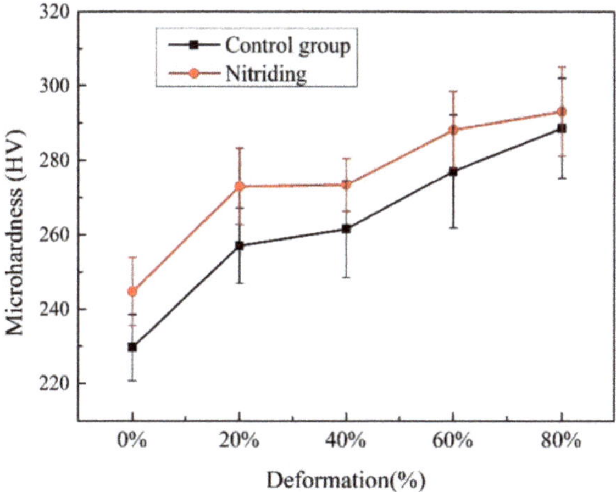

Figure 12. Microhardness values of cold-rolled deformed alloy samples before and after nitriding.

4. Conclusions

1. As the cold-rolling deformation rate of TA2 alloy increases, obvious grain refinement occurs, and the grain orientation changes from TD to RD and then to TD. The {11-20} cylindrical texture component reduces, while the {0001} basal surface texture and {10-10} cone texture component increase, and the cone texture component is the largest.
2. It can be seen from the grain characteristics that the deformed sample undergoes recrystallization during low-temperature nitriding. The higher the degree of cold-rolling deformation, the higher the degree of recrystallization, and the higher the degree of microstructure refinement. The comparison of texture characteristics before and after nitriding shows that the direction of {0001} basal texture is deflected from TD to RD, and {10-10} cylindrical texture components gradually increase; the c-axis of the TD-ND plane has 45–90° deflection in the ND direction. As the degree of recrystallization increases, the texture orientation of the special angle gradually disappears.
3. After low-temperature plasma nitriding of the deformed alloy sample, the metastable nitriding phase TiN0.26 is formed on the surface. The surface N element content increases first and then decreases; this is because as the degree of deformation increases, the bonding force between atoms in the alloy matrix decreases and the defect density increases. Under different deformations, no obvious nitride layer is formed; this study may have formed a new type of nitrided zone. The composite process described in this article can significantly improve the wear resistance and hardness of deformed alloy samples.
4. In this paper, the cold-rolling deformation and low-temperature plasma nitriding caused simultaneous recrystallization and composite modification processes that we can use to effectively achieve the dual goals of alloy matrix structure control and surface modification and is of great significance for obtaining TA2 alloy materials with a strong plastic ratio.

Author Contributions: Conceptualization, G.L. and H.S.; methodology, E.W.; software, K.S.; validation, X.Z. and Y.F.; formal analysis, G.L., H.S., and K.S.; investigation, K.S.; resources, Y.F. and X.Z.; data curation, E.W.; writing—original draft preparation, G.L.; writing—review and editing, G.L.; visualization, Y.F.; supervision, Y.F.; project administration, Y.F. All authors have read and agreed to the published version of the manuscript.

Funding: This research was funded by the Development of high-strength, high-toughness and large-size aluminum alloy plates for domestic large passenger aircraft, the type is Heilongjiang Province's "Hundreds, Millions" Engineering Science and Technology Major Project. This research was also funded by the Formation Mechanism and Control Technology of Functionalized Modified Coating on Titanium Alloy Surface, grant number 61429090307.

Institutional Review Board Statement: Not applicable.

Informed Consent Statement: Not applicable.

Data Availability Statement: Data is contained within the article.

Conflicts of Interest: Authors declare there are no conflict of interest.

References

1. Rack, H.J.; Qazi, J.I. Titanium alloys for biomedical applications. *Mater. Sci. Eng. C* **2006**, *26*, 1269–1277. [CrossRef]
2. Calvert, E.L.; Knowles, A.J.; Pope, J.J.; Dye, D.; Jackson, M. Novel high strength titanium-titanium composites produced using field-assisted sintering technology (FAST). *Scr. Mater.* **2019**, *159*, 51–57. [CrossRef]
3. Singh, P.; Pungotra, H.; Kalsi, N.S. On the characteristics of titanium alloys for the aircraft applications. *Mater. Today Proc.* **2017**, *4*, 8971–8982. [CrossRef]
4. Leyens, C.; Peters, M. (Eds.) *Titanium and Titanium Alloys: Fundamentals and Applications*; Wiley-VCH GmbH: Berlin, Germany, 2003.
5. Wang, X.; Qu, Z.; Li, J.; Zhang, E. Comparison study on the solution-based surface biomodification of titanium: Surface characteristics and cell biocompatibility. *Surf. Coat. Technol.* **2017**, *329*, 109–119. [CrossRef]
6. Lütjering, G.; Williams, J.C. *Titanium Based Intermetallics*; Springer: Berlin/Heidelberg, Germany, 2003.
7. Ruiliang, L.; Mufu, Y.; Yingjie, Q.; Yudong, F. Low temperature plasma RE nitrocarburizing of martensitic stainless steel. *Heat Treat. Met.* **2013**, *38*, 54–58.
8. Suwas, S.; Beausir, B.; Tóth, L.S.; Fundenberger, J.J.; Gottstein, G. Texture evolution in commercially pure titanium after warm equal channel angular extrusion. *Acta Mater.* **2011**, *59*, 1121–1133. [CrossRef]
9. Wang, Y.; He, W.; Liu, N.; Chapuis, A.; Luan, B.; Liu, Q. Effect of pre-annealing deformation on the recrystallized texture and grain boundary misorientation in commercial pure titanium. *Mater. Charact.* **2017**, *136*, 1–11. [CrossRef]
10. Gurao, N.P.; Sethuraman, S.; Suwas, S. Evolution of Texture and Microstructure in Commercially Pure Titanium with Change in Strain Path during Rolling. *Metall. Mater. Trans. A* **2013**, *44*, 1497–1507. [CrossRef]
11. Sahoo, S.K.; Sabat, R.K.; Sahni, S.; Suwas, S. Texture and microstructure evolution of commercially pure titanium during hot rolling: Role of strain-paths. *Mater. Des.* **2016**, *91*, 58–71. [CrossRef]
12. Bishoyi, B.D.; Sabat, R.K.; Sahoo, S.K. Effect of temperature on microstructure and texture evolutions during uniaxial compression of commercially pure titanium. *Mater. Sci. Eng. A* **2018**, *718*, 398–411. [CrossRef]
13. Rollett, A. *Recrystallization and Related Annealing Phenomena*; Elsevier: Amsterdam, The Netherlands, 1995.
14. Luo, Z.P.; Guo, X.K.; Hou, J.X.; Zhou, X.; Li, X.Y.; Lu, K. Plastic deformation induced hexagonal-close-packed nickel nano-grains. *Scr. Mater.* **2019**, *168*, 67–70. [CrossRef]
15. Zhang, X.Y.; Li, B.; Liu, Q. Non-equilibrium basal stacking faults in hexagonal close-packed metals. *Acta Mater.* **2015**, *90*, 140–150. [CrossRef]
16. Wang, Z.; Cochrane, C.; Skippon, T.; Dong, Q.; Daymond, M.R. Dislocation evolution at a crack-tip in a hexagonal close packed metal under plane-stress conditions. *Acta Mater.* **2018**, *164*, 25–38. [CrossRef]
17. Kaita, W.; Hagihara, K.; Rocha, L.A.; Nakano, T. Plastic deformation mechanisms of biomedical Co–Cr–Mo alloy single crystals with hexagonal close-packed structure. *Scr. Mater.* **2017**, *142*, 111–115. [CrossRef]
18. Contieri, R.J.; Zanotello, M.; Caram, R. Recrystallization and grain growth in highly cold worked CP-Titanium. *Mater. Sci. Eng. A* **2010**, *527*, 3994–4000. [CrossRef]
19. Hayama, A.O.; Andrade, P.N.; Cremasco, A.; Contieri, R.J.; Afonso, C.R.; Caram, R. Effects of composition and heat treatment on the mechanical behavior of Ti–Cu alloys. *Mater. Des.* **2014**, *55*, 1006–1013. [CrossRef]
20. Hayama, A.O.; Lopes, J.F.; Da Silva, M.J.; Abreu, H.F.; Caram, R. Crystallographic texture evolution in Ti–35Nb alloy deformed by cold rolling. *Mater. Des.* **2014**, *60*, 653–660. [CrossRef]
21. Kovacı, H.; Hacısalihoğlu, İ.; Yetim, A.F.; Çelik, A. Effects of shot peening pre-treatment and plasma nitriding parameters on the structural, mechanical and tribological properties of AISI 4140 low-alloy steel. *Surf. Coat. Technol.* **2019**, *358*, 256–265. [CrossRef]

22. Kovacı, H.; Bozkurt, Y.B.; Yetim, A.F.; Aslan, M.; Çelik, A. The effect of surface plastic deformation produced by shot peening on corrosion behavior of a low-alloy steel. *Surf. Coat. Technol.* **2019**, *360*, 78–86. [CrossRef]
23. Xing'an, W.; Mufu, Y.; Ruiliang, L.; Zhang, Y. Effect of rare earth addition on microstructure and corrosion behavior of plasma nitrocarburized M50NiL steel. *J. Rare Earths* **2016**, *11*, 86–93.
24. Sun, H.Z.; Zheng, J.; Song, Y.; Chi, J.; Fu, Y.D. Effect of the deformation on nitrocarburizing microstructure of the cold deformed Ti-6Al-4V alloy. *Surf. Coat. Technol.* **2019**, *362*, 234–238. [CrossRef]
25. Zhu, X.S.; Fu, Y.D.; Li, Z.F.; Leng, K. Wear resistance of TC4 by deformation accelerated plasma nitriding at 400 °C. *J. Cent. South Univ.* **2016**, *23*, 2771–2776. [CrossRef]
26. Fu, Y.D.; Zhu, X.S.; Li, Z.F.; Ke, L.E. Properties and microstructure of Ti6Al4V by deformation accelerated low temperature plasma nitriding. *Trans. Nonferr. Met. Soc. China* **2016**, *26*, 2609–2616. [CrossRef]
27. Kikuchi, S.; Ota, S.; Akebono, H.; Omiya, M.; Komotori, J.; Sugeta, A.; Nakai, Y. Formation of nitrided layer using atmospheric-controlled IH-FPP and its effect on the fatigue properties of Ti-6Al-4V alloy under four-point bending. *Procedia Struct. Integr.* **2016**, *2*, 3432–3438. [CrossRef]
28. Tokaji, K.; Ogawa, T.; Shibata, H. The effects of gas nitriding on fatigue behavior in titanium and titanium alloys. *J. Mater. Eng. Perform.* **1999**, *8*, 159–167. [CrossRef]
29. Braz, J.K.; Martins, G.M.; Morales, N.; Naulin, P.; Fuentes, C.; Barrera, N.P.; Vitoriano, J.O.; Rocha, H.A.; Oliveira, M.F.; Alves, C., Jr.; et al. Live endothelial cells on plasma-nitrided and oxidized titanium: An approach for evaluating biocompatibility. *Mater. Sci. Eng. C* **2020**, *113*, 111014. [CrossRef] [PubMed]
30. Zeng, C.; Wen, H.; Ettefagh, A.H.; Zhang, B.; Gao, J.; Haghshenas, A.; Raush, J.R.; Guo, S.M. Laser nitriding of titanium surfaces for biomedical applications. *Surf. Coat. Technol.* **2020**, *385*, 125397. [CrossRef]

coatings

Article

Low-Temperature Plasma Nitriding of 3Cr13 Steel Accelerated by Rare-Earth Block

Yuan You [1,2,3,*], Rui Li [1], Mufu Yan [2], Jihong Yan [3], Hongtao Chen [4], Chaohui Wang [1], Dongjing Liu [1], Lin Hong [1] and Tingjie Han [1]

1. School of Materials Science and Engineering, Qiqihar University, Qiqihar 161006, China; ly731374908@163.com (R.L.); wch800209@126.com (C.W.); lovccat@163.com (D.L.); hl25802021@163.com (L.H.); htj202021@163.com (T.H.)
2. National Key Laboratory of Metal Precision Thermal Processing, School of Materials Science and Engineering, Harbin Institute of Technology, Harbin 150001, China; yanmufu@hit.edu.cn
3. School of Mechanical and Electrical Engineering, Harbin Institute of Technology, Harbin 150001, China; jyan@hit.edu.cn
4. School of Materials Science and Engineering, Harbin University of Science and Technology, Harbin 150040, China; htchen83@163.com
* Correspondence: youyuan@qqhru.edu.cn

Abstract: The plasma nitriding of 3Cr13 steel occurred at 450 °C for 4, 8 and 12 h in NH_3 with and without rare earth (RE). The nitrided layers were characterized using an OM, SEM, TEM, XRD, XPS, microhardness tester and electrochemical workstation. The modified layer, with and without La, are composed of a compound layer and diffusion layer from surface to core. After the addition of La during nitriding, the maximum increase of layer thickness, mass gain and average microhardness was 15.6%, 35.8% and 212.50$HV_{0.05}$, respectively. With the increase of the proportion of ε-$Fe_{2-3}N$, the passivation zone of the corrosion resistance curve increases from 2.436 to 3.969 V, the corrosion current density decreases, the corrosion potential and pitting potential both increase, and, consequently, the corrosion resistance is significantly improved. Most of the surface microstructures of the nitrided layer was refined by the addition of La. The presence of La reduces the N content in the modified layer, which accelerates the diffusion of N atoms and, thus, accelerates the nitriding process.

Keywords: 3Cr13 steel; plasma nitriding; microstructure; microhardness; corrosion resistance

Citation: You, Y.; Li, R.; Yan, M.; Yan, J.; Chen, H.; Wang, C.; Liu, D.; Hong, L.; Han, T. Low-Temperature Plasma Nitriding of 3Cr13 Steel Accelerated by Rare-Earth Block. *Coatings* **2021**, *11*, 1050. https://doi.org/10.3390/coatings11091050

Academic Editor: Alexander D. Modestov

Received: 29 July 2021
Accepted: 28 August 2021
Published: 31 August 2021

Publisher's Note: MDPI stays neutral with regard to jurisdictional claims in published maps and institutional affiliations.

Copyright: © 2021 by the authors. Licensee MDPI, Basel, Switzerland. This article is an open access article distributed under the terms and conditions of the Creative Commons Attribution (CC BY) license (https:// creativecommons.org/licenses/by/ 4.0/).

1. Introduction

Because of its low price, it is suitable for a stainless steel working environment with general requirements. Because of its impact resistance, processing formability and high plasticity, it is widely used to manufacture ice blades, scalpels, piston rods, turbine blades and valve parts [1,2]. It can be used for surgical cutting tools, tool materials and structural materials in the fields of medical instruments, the plastics industry and mechanical construction [3]. Compared with other stainless steels, 3Cr13 steel has enhanced toughness and reduced hardness, with relatively worse rust resistance. In order to improve its service life and expand its application, researchers have investigated techniques such as salt bath nitriding [4], cathode cage nitriding [5], radio frequency magnetron sputtering nitriding [6], active screen plasma nitriding [7] and pulse DC plasma nitriding [8] technology processing.

From an economic point of view, it not only reduces the cost, but also improves the surface properties of various steels in an environmentally friendly way. As an efficient and energy-saving thermochemical surface processing and modification technology, it has the advantages of a short processing time, low temperature, reduced brittleness of the nitriding layer, a reduction of work piece deformation and surface flexibility and toughness [9].

China possesses the world's largest reserve of RE elements, which play an important role in the metallurgical industry, military, petrochemical, glass, ceramics, new materials and agriculture. RE elements are mainly used in tanks, aircraft, lighting sources, polishes and exhaust purification catalysts [10].

In recent years, scientists have explored the impacts of adding RE elements to steel in terms of corrosion resistance [11], strengthening effect [12], contact fatigue limit, bending fatigue limit [13] and other mechanical properties of steel [14,15]. The effects of RE types (cast lanthanum La, pure La block, La_2O_3, Er_2O_3, Yb_2O_3, CeO_2) on the strength, corrosion resistance and fatigue life of steel were also studied. Zhang [16] et al. proved that La reduced the N content in the nitrided layer and promoted the diffusion of interstitial nitrogen from the surface to the interior of the workpiece. La atoms can also attract each other with N atoms in order to achieve the ability of denitrification.

However, there are few studies on plasma nitriding of 3Cr13 stainless steel with the addition of RE. Therefore, it is necessary to conduct a detailed analysis of the com properties of 3Cr13 stainless steel with and without the addition of RE. In this paper, 3Cr13 steel has been plasma nitrided at 450 °C for 4, 8 and 12 h in NH_3, with and without RE. The effects of RE (La) on the microstructure, phase composition, microhardness and corrosion resistance of the modified layer were studied.

2. Materials and Methods

The material used in the present work is 3Cr13 steel with the following chemical composition (wt.%): 0.33C, 12.6Cr, 0.1Ni, 0.48Si, 0.52P, 0.01S and balance Fe. Before nitriding, the steel was solution treated in the box electric furnace SXL-1400C (Shanghai Jvjing Precision Instrument Manufacturing Co., Ltd., Shanghai, China) at 960 °C for 1 h and then quenched in oil. The surface of the sample was polished with water sandpaper (#120 and #240), and then cleaned with alcohol by ultrasonication.

Before using the LDMC-30AFZ plasma nitriding furnace (Wuhan Shoufa Surface Engineering Co., Ltd., Wuhan, China) for the plasma nitriding treatment, the chamber was evacuated to below 10 Pa by a rotary pump. The RE used in the experiment came from the RE block with a volume of $(0.5 \times 0.5 \times 1)$ cm^3, namely RE0.25. The samples were uniformly tied to the sample frame with iron wire, and the RE block was tied to the center of the sample frame. The nitriding conditions are the same with or without RE. The nitriding experiment was conducted in ammonia (NH_3) with a flux of 100 mL/min at a temperature of 450 °C for 4, 8 and 12 h, and the voltage and working current during nitriding were 260 Pa and 12 A, respectively. After plasma nitriding, the specimens were cooled down slowly inside the vacuum furnace.

The weight gain during the experimental process was obtained using CPA-225D electronic balance (Sai Dolis scientific instruments Beijing Co., Ltd, Beijing, China) with an accuracy of 0.00001 g. A metallographic samples inlaying machine (XQ-1, Shanghai Metallurgical Equipment Company Ltd., Shanghai, China) was used for cross-section sample preparation. The models of grinding water sandpaper were 120#, 240#, 400#, 600#, 800#, 1000#, 1200#, 1500# and 2000#, respectively. The polished samples were etched using a marble solution (2 g $CuSO_4$ + 10 mL HCl + 50 mL H_2O) for 1 s. The cross-section micrographs were observed using an optical microscope (OM, 9XB-PC, Shanghai Optical Instrument No.1 Factory, Shanghai, China). Microhardness was measured with a microhardness tester (HV-1000IS, Shanghai Jvjing Precision Instrument Manufacturing Co., Ltd., Shanghai, China) under an indentation load of 50 g for 15 s. The anodic polarization tests were employed to estimate the corrosion resistance of the specimens in a 3.5 wt.% NaCl solution, the reference electrode was Ag/AgCl and a platinum column was used as the auxiliary electrode. The anodic polarization curves of the specimens were recorded at a sweeping speed of 0.01 V/s. X-ray diffraction (XRD) with Cu-Kα radiation (λ = 0.15418 nm) was carried out in the range of angles 20°–100° at 45 kV and 200 mA with a 5° interval step mode using a D8 Advance diffractometer (Bruker, Billerica, MA, USA). The cross-section morphologies of the modified surfaces and the changes in element content with depth were analyzed using a S-3400 scanning electron microscope (SEM, Hitachi, Tokyo, Japan), coupled with energy dispersive X-ray analyzer (EDS) facilities. XPS analysis was performed in an ultra-high vacuum using a Escalab250XI (Thermo Fisher Scientific, Waltham, MA, USA). The excitation source was monochromatic Al Kα radiation, operated at 15 kV and the emission current was 10 mA. The binding energy of C1s neutral carbon peak at 284.6 eV was taken as the reference value and the measured binding energy was modified. The argon ion spray gun (kinetic energy of 3 keV) was used to complete the 30 min incident argon ion sputtering process on the samples in the preparation chamber to ensure that the surface of the samples, at a depth of 100 nm, was clean. The microstructure and grain size of the amorphous samples were observed using a transmission electron microscopy (TEM, JEOL Ltd., Tokyo, Japan) at the operating voltage of 300 kV.

3. Results and Discussion

3.1. Cross-Sectional Microstructure and Depth Analysis

Figure 1a,b are cross-sectional micrographs of 3Cr13 steel after plasma nitriding for 4, 8 and 12 h without and with RE. The nitrided layer is composed of a compound layer and diffusion layer from surface to core. There are distinct boundaries between the compound layer, the diffusion layer and the substrate. The total thickness of the nitrided layer can be obtained according to the microhardness diagrams, the thickness of the compound layer can be obtained according to the ruler in Figure 1, and the difference value is the thickness of the diffusion layer. With the addition of RE, the thickness of the nitrided layer increases with time, and the total thickness of the nitrided layer is 130.4, 161.5 and 185.6 μm, respectively, which is 8.6%, 15.6% and 14.7% higher than that without RE, respectively.

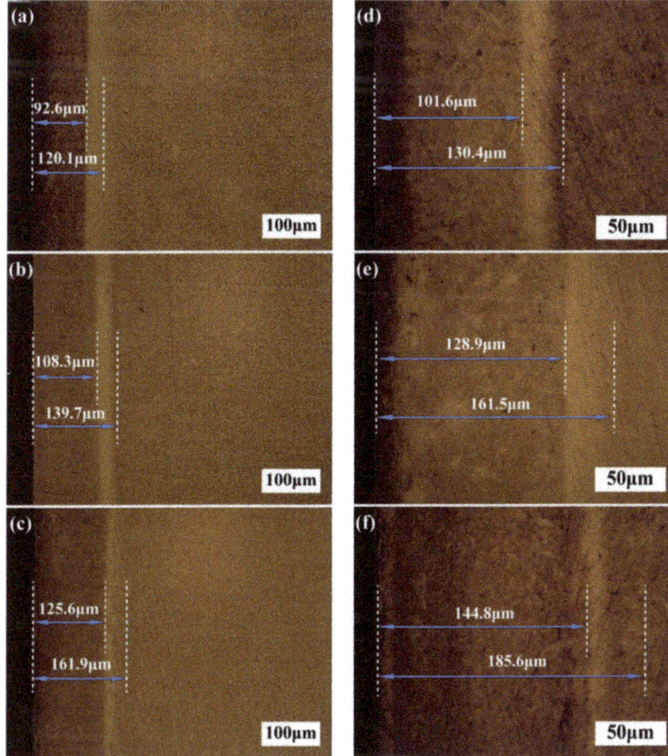

Figure 1. Cross-section micrograph of modified layer of 3Cr13 steel after plasma nitriding (**a**) h4, (**b**) h8, (**c**) h12 without RE and (**d**) h4, (**e**) h8, (**f**) h12 with RE.

Figure 2a,b are histograms of the thickness increase of the modified layer section of 3Cr13 steel after plasma nitriding for 4, 8 and 12 h without and with RE. The change in the thickness of the nitrided layer can be seen more intuitively.

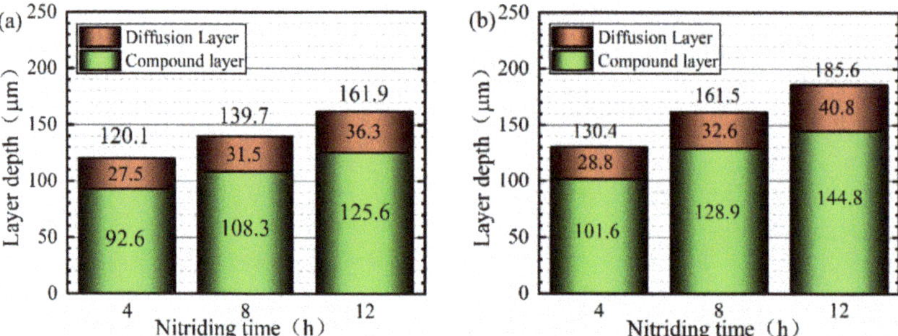

Figure 2. Thickness of modified layer section of 3Cr13 steel after plasma nitriding at different times (**a**) Without RE (**b**) With RE.

Figure 3a,b are histograms of the weight gain of 3Cr13 steel after plasma nitriding for 4, 8 and 12 h without and with RE. The weight gain of the sample becomes more significant with the extension of time. With the addition of RE, the weight gain of the nitrided layer is very obvious, reaching 2.31, 3.56 and 4.23 mg/cm^2, respectively, which is 102.6%, 135.8% and 121.5% higher than that without the addition of RE, respectively.

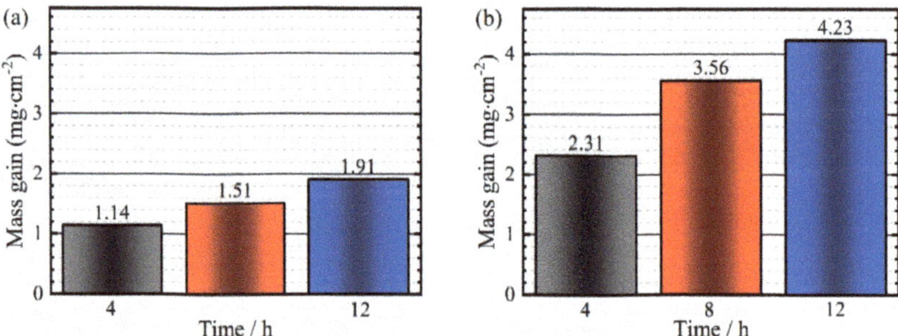

Figure 3. Mass increase per unit area of 3Cr13 steel after plasma nitriding at different times (**a**) Without RE (**b**) With RE.

3.2. Microhardness Profile

Figure 4a,b are the microhardness diagrams of 3Cr13 steel after plasma nitriding for 4, 8 and 12 h without and with RE. Microhardness varies with the nitriding time and depth of the nitriding layer. From the surface layer to the dense compound layer, the microhardness first increases (gradual horizontal hardening), reaches the peak and then gradually decreases. A platform appears (flat horizontal hardening) when it reaches the diffusion layer. There is a sudden drop from the diffusion layer to the substrate boundary and then a gradual transition to the substrate. After 4, 8 and 12 h of nitriding, the average hardness of the modified layer is 813.53HV$_{0.05}$. After 4, 8 and 12 h of RE nitriding, the average hardness of the modified layer is 1026.03HV$_{0.05}$, which demonstrates an increase of 212.50HV$_{0.05}$ compared to the sample without RE. The increase in surface hardness was directly related to the formation of fine and uniform ε-Fe$_{2-3}$N and γ'-Fe$_4$N nitride phases in the cemented layer, as shown in Figure 5. The comparison between the microhardness and the depth profile of nitrogen concentration indicated that this was related to the diffusion of N. The N content on the surface (0 µm) reaches the maximum and the surface hardness reaches the maximum. From the compound layer to the diffusion layer and until the substrate, the surface hardness decreased significantly with the decrease of N content.

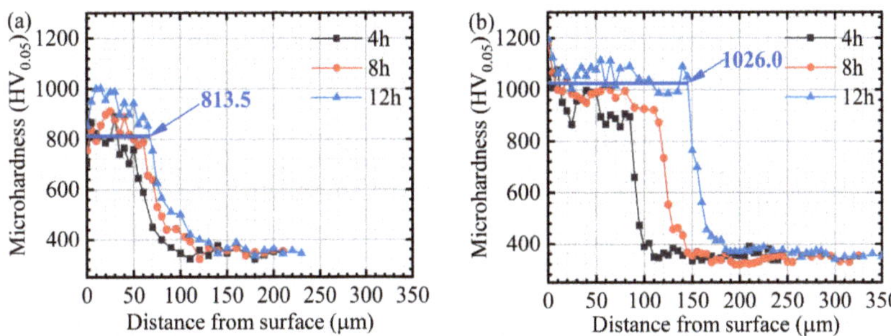

Figure 4. The microhardness diagrams of 3Cr13 steel after plasma nitriding at different times (**a**) Without RE (**b**) With RE.

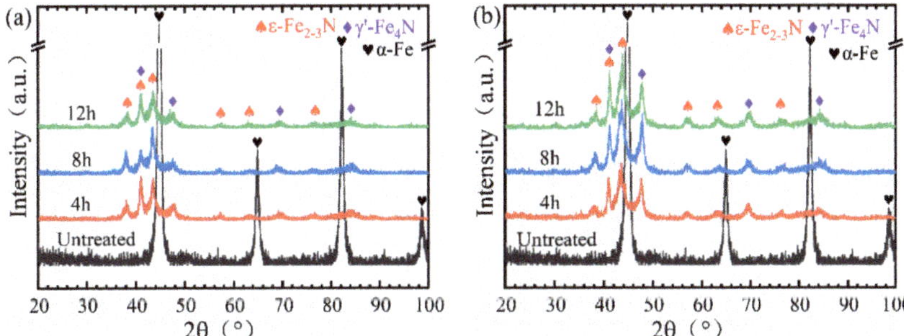

Figure 5. XRD pattern of the modified layer of 3Cr13 steel after plasma nitriding at different times (**a**) Without RE (**b**) With RE.

3.3. XRD

Figure 5a,b are the XRD pattern of the modified layer of 3Cr13 steel after plasma nitriding for 4, 8 and 12 h without and with RE. In Figure 5a, the main phase is ε-Fe$_{2\text{-}3}$N, the secondary phase is γ'-Fe$_4$N. After the addition of La, the intensity of the ε-Fe$_{2\text{-}3}$N diffraction peak is significantly enhanced, and the content of ε-Fe$_{2\text{-}3}$N in the modified layer is greatly increased. Reference [14] shows that ε-Fe$_{2\text{-}3}$N improved the corrosion resistance to a significant extent, and more than that of γ'-Fe$_4$N. At the same time, due to the La atoms dissolving into the nitride layer, the lattice distortion results in increased microhardness [12].

3.4. Polarization Curve and Fitting Data Analysis

Figure 6a,b display the corrosion resistance curve of 3Cr13 steel after plasma nitriding for 4, 8 and 12 h without and with RE. With or without the addition of RE, the passivation zone of the corrosion resistance curve becomes wider with the increase of time, the corrosion resistance is slightly enhanced.

The data relating to the corrosion rate, corrosion potential (E_0), polarization resistance (Rp) and corrosion current (I_0) on the surface of the modified layer are shown in Tables 1 and 2. In comparing the samples with and without RE nitriding, with the increase of time, the corrosion rate of the modified layer decreased slightly, and the corrosion resistance of the modified layer increased slightly. It can be concluded that the modified layer has the best corrosion resistance after 12 h of RE nitriding, with the corrosion rate decreasing to 0.9255×10^{-2} mm/a.

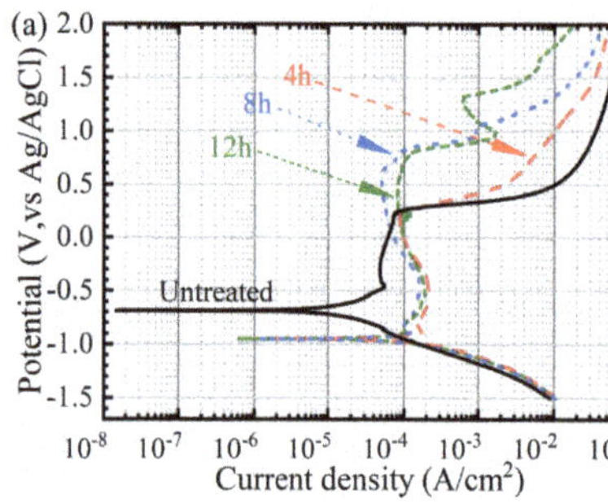

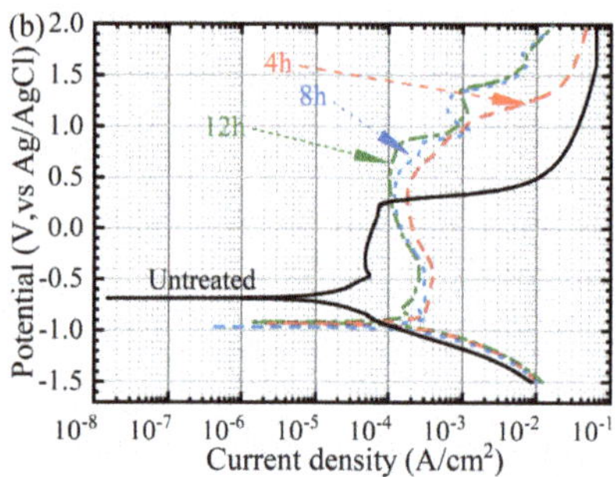

Figure 6. Corrosion resistance curve of modified layer of 3Cr13 steel after plasma nitriding at different times (**a**) Without RE (**b**) With RE.

Table 1. Fitting data of the polarization curve of the modified layer of 3Cr13 steel after plasma nitriding at different times.

Time/h	Corrosion Rate/($\times 10^{-2}$ mm·a^{-1})	R_p/($\times 10^{-3}$ Ω·cm^{-2})	I_0/($\times 10^{-5}$ A·cm^{-2})	E_0/V	Passivation Zone/V
Untreated	49.746	61.681	4.22930	−0.87501	1.038
4	1.3519	2.2697	1.14930	−0.94865	1.352
8	1.2851	2.3995	1.04978	−0.94805	1.801
12	1.0895	2.7796	1.00630	−0.95095	2.436

Table 2. Fitting data of the polarization curve of the modified layer of 3Cr13 steel after RE plasma nitriding at different times.

Time/h	Corrosion Rate/($\times 10^{-2}$ mm·a^{-1})	R_p/($\times 10^{-3}$ Ω·cm^{-2})	I_0/($\times 10^{-5}$ A·cm^{-2})	E_0/V	Passivation Zone/V
Untreated	49.746	61.681	4.22930	−0.87501	1.038
4	1.2288	2.3767	0.89491	−0.91983	2.105
8	1.0463	2.4935	0.73978	−0.94882	2.401
12	0.9255	2.9950	0.63733	−0.92949	2.563

The enhanced corrosion resistance is due to the presence of a dense nitride-rich layer on the ε-nitride surface, as more nitride helps protect the surface from corrosion [17]. Figure 5 proved that the formation of ε-Fe$_{2-3}$N and γ'-Fe$_4$N greatly improves the corrosion resistance of the modified layer [18].

3.5. SEM

The C, N, Cr Fe and La elements at different locations of the modified layer were measured by point scanning. The content of N element decreases gradually from the surface to the substrate, while the content of Fe element increases gradually.

As shown in Figure 7a, as the depth of the nitriding layer increases, Fe content gradually decreases, N content reaches as high as 4.57 at.%, and then continues to decline to 0.23 at.%. As shown in Figure 7b, as the depth of the nitriding layer increases, Fe content gradually decreases, the N content in the surface layer does not reach the maximum, but gradually increased to a maximum of 2.72%, and then gradually transitioned to the substrate.

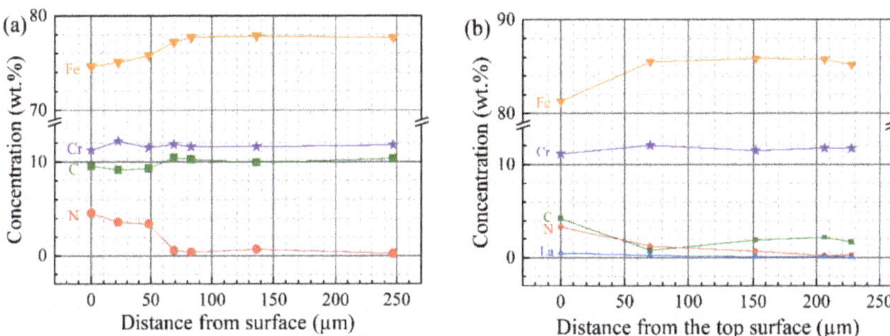

Figure 7. Element content depth distribution of the modified layer of 3Cr13 steel after plasma nitriding for 12 h (**a**) Without RE (**b**) With RE.

First, ion bombardment produces La ions and some neutral ions, which are deposited on the surface of stainless steel and play a role in cleaning and activation, increasing the concentration of surface nitrogen [9]. Then, due to the large size difference between La and Fe atoms, La causes obvious distortion in the surrounding lattice. Meanwhile, La enhances the surface bombardment effect, increasing the crystal defects such as vacancies and dislocations and further reducing the N content in the nitrided layer, which accelerates the diffusion of N atoms and produces solid solution strengthening. Thus, the thickness of compound and effective hardening layer is increased, and the nitriding process is accelerated [19,20].

Figure 8a,b are the element depth profile of the modified layer of 3Cr13 steel after plasma nitriding for 12 h, without and with RE. The percentage of La, C and N are measured by linear scanning. As can be seen from Figure 8a,b, the Cr content of both C-layer (0–125.6 μm) and D-layer (0–144.8 μm) increased, while the N content decreased significantly. As shown in Figure 9a, the total diffusion depth increased from 145.3 to 157.9 μm, the contents of Cr and N decreased in the D-layer layer (125.6–145.3 μm), while the contents of Cr and N basically remained unchanged in the D-layer (144.8–157.9 μm) layer in Figure 9b. The La content decreased gradually. The addition of La greatly promoted the diffusion of N and reduced the precipitation of CrN.

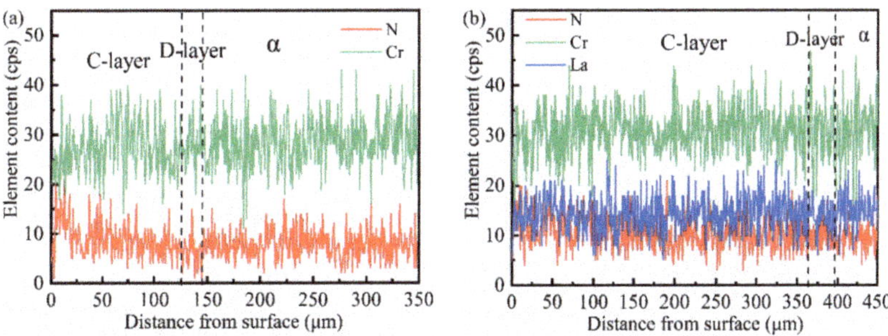

Figure 8. Element depth profile of modified layer of 3Cr13 steel after plasma nitriding for 12 h (**a**) Without RE (**b**) With RE.

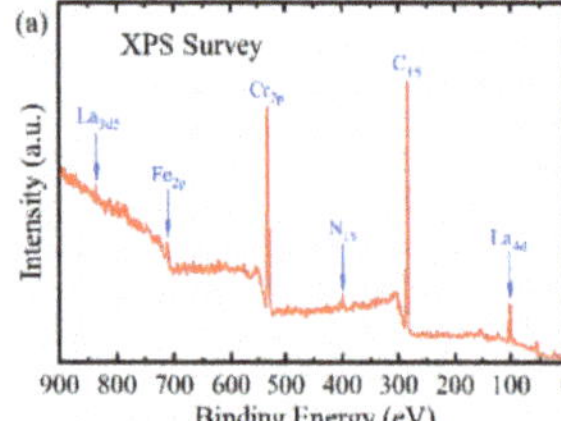
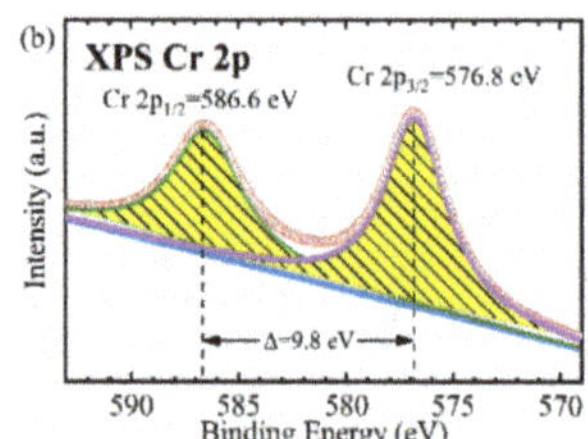
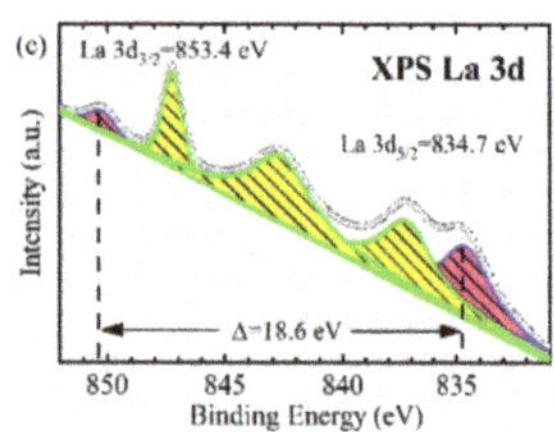

Figure 9. XPS spectra modified layer of 3Cr13 steel after RE plasma nitriding for 12 h (**a**) Survey spectra (**b**) Cr 2p spectra (**c**) La3d spectra.

3.6. XPS

The detailed chemical characteristics of various elements in the surface layer are analyzed by XPS characterization in order to obtain information about the formation of surface nitride in the plasma ionitriding process. The results are shown in Figure 9a.

In order to determine the possibility of producing a Cr-N bond on the sample surface under consideration and to study the content of CrN on the surface, the XPS spectrum of a high-resolution Cr2P is shown in Figure 9b. A strong peak (a weak peak) was observed in the XPS spectrum of Cr2P, with a peak position matching the orbital 2P1/2 (586.6 eV) and orbital 2P2/3(576.8 eV) [21].

Figure 9c shows the photoemission spectra of La 3D rays obtained by surface modification with monochromatic Al Ka photons. The main peak of the La3d XPS spectrum was about 853.4 and 834.7 eV, belonging to La3d3/2 and La3d5/2, respectively [22]. The results show that the successful diffusion of the La element to the nitride surface can not only spread to a significant depth of the surface, but can also help the gap nitrogen to diffuse to a deeper layer of the workpiece, and improve the structure and properties of the modified layer.

3.7. TEM

Figure 10a,b show typical transmission electron microscope images and diffraction ring images of nanocrystallization of 3Cr13 steel at 10 μm from the sample surface after RE nitriding for 12 h. According to the scale and the comparison of the standard XRD diffraction peak pattern in Figure 10b, the inner ring corresponds to the crystal plane γ'-Fe$_4$N (311), and the outer ring corresponds to the crystal plane ε-Fe$_3$N (112). The SAED patterns shows that there are a large number of γ'-Fe$_4$N and ε-Fe$_3$N particles. The addition of La can further refine the grain size of the nanocrystalline layer [23], and the results indicate that the addition of La can produce a microalloying effect on the nitride layer of 3Cr13 steel.

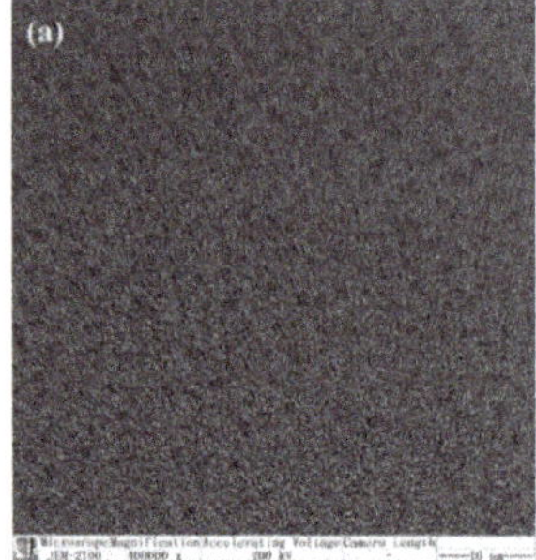
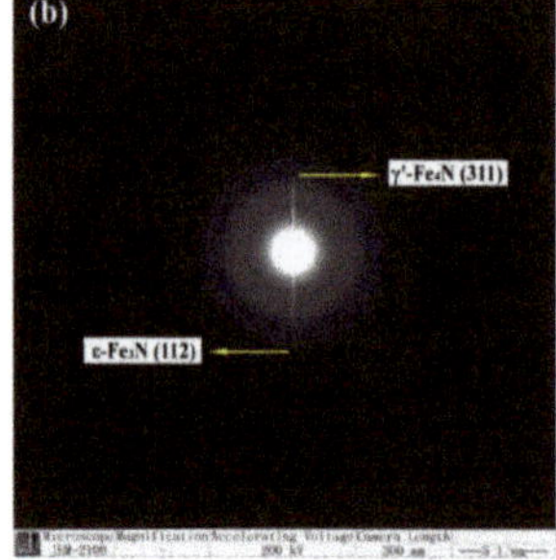

Figure 10. TEM bright field image (**a**) and corresponding SAED pattern (**b**) of surface nanometer layer of modified 3Cr13 steel after RE plasma nitriding for 12 h.

4. Conclusions

In this paper, 3Cr13 steel was plasma nitrided for 4, 8 and 12 h at 450°C in NH3 atmosphere with and without RE, and the following conclusions were drawn:

1. The nitrided layer was composed of a compound layer and diffusion layer from surface to core. The total thickness of the RE nitrided layer increased by 8.6%, 15.6% and 14.7% more than that of the layer without RE. The thickness of the compound layer was increased by 9.7%, 19.0% and 15.3% more than that of the compound layer without RE. The mass gain of the RE nitrided layer was significant, and increased by 36% more than that of plain nitriding.
2. The modified layer was mainly composed of ε-$Fe_{2.3}N$ and γ'-Fe_4N nitride phases. The addition of RE (La) increased the phase ratio of ε-$Fe_{2.3}N$ in the surface layer. The corrosion current density was decreased, the corrosion resistance was increased, and the passivation zone gradually widened. The addition of RE (La) made the passivation more obvious, thus improving the corrosion resistance.
3. The average hardness of the compound layer was $813.53HV_{0.05}$. The average hardness of the RE compound layer was $1026.03HV_{0.05}$, which increased by $212.50HV_{0.05}$.
4. SEM (EDS) and XPS demonstrated the presence of La, and the N content gradually decreased with the depth of the nitride layer. La diffused into the interior of the layers and further reduced the N content in the nitride layer, which accelerated the diffusion of N atoms and thus promoted the nitriding process.
5. There are many γ'-Fe_4N and ε-Fe_3N particles in the nitrided layer. The addition of RE further refined the microstructure in the nitrided layer, in which the microalloying effect was produced to strengthen the modified layer.

Author Contributions: Investigation, R.L. and Y.Y.; Writing—Review and Editing, R.L., Y.Y., M.Y., J.Y., H.C., C.W., D.L., L.H. and T.H. All authors have read and agreed to the published version of the manuscript.

Funding: Supported by the National Natural Science Foundation of China (51401113) and Natural Science Foundation of Heilongjiang Province of China (E2016069), Heilongjiang Postdoctoral Financial Assistance (LBH-Z16061), the Fundamental Research Funds in Heilongjiang Provincial Universities (135309504).

Institutional Review Board Statement: Not applicable.

Informed Consent Statement: Not applicable.

Data Availability Statement: Data is contained within the article.

Conflicts of Interest: The authors declare no conflict of interest.

References

1. Xi, Y.T.; Liu, D.X.; Han, D. Improvement of corrosion and wear resistances of AISI 420 martensitic stainless steel using plasma nitriding at low temperature. *Surf. Coat. Technol.* **2008**, *202*, 2577–2583. [CrossRef]
2. Pinedo, C.E.; Monteiro, W.A. On the kinetics of plasma nitriding a martensitic stainless steel type AISI 420. *Surf. Coat. Technol.* **2004**, *179*, 119–123. [CrossRef]
3. Çetin, A.; Tek, Z.; Öztarhan, A.; Artunc, N. A comparative study of single and duplex treatment of martensitic AISI 420 stainless steel using plasma nitriding and plasma nitriding-plus-nitrogen ion implantation techniques. *Surf. Coat. Technol.* **2007**, *2019*, 8127–8130. [CrossRef]
4. Luo, G.; Zheng, Z.; Ning, L.; Tan, Z.; Tong, J.; Liu, E. Failure analysis of AISI 316L ball valves by salt bath nitriding. *Eng. Fail. Anal.* **2020**, *111*, 104455. [CrossRef]
5. Shen, H.; Wang, L. Mechanism and properties of plasma nitriding AISI 420 stainless steel at low temperature and anodic (ground) potential. *Surf. Coat. Technol.* **2020**, *403*, 126390. [CrossRef]
6. Nakasa, K.; Gao, S.; Yamamoto, A.; Sumomogi, T. Plasma nitriding of cone-shaped protrusions formed by sputter etching of AISI 420 stainless steel and their application to impression punch to form micro-holes on polymer sheets. *Surf. Coat. Technol.* **2019**, *358*, 891–899. [CrossRef]
7. Li, Y.; He, Y.; Xiu, J.; Wang, W.; Zhu, Y.; Hu, B. Wear and corrosion properties of AISI 420 martensitic stainless steel treated by active screen plasma nitriding. *Surf. Coat. Technol.* **2017**, *329*, 184–192. [CrossRef]

8. Corengia, P.; Ybarra, G.; Moina, C.; Cabo, A.; Broitman, E. Microstructural and topographical studies of DC-pulsed plasma nitrided AISI 4140 low-alloy steel. *Surf. Coat. Technol.* **2005**, *200*, 2391–2397. [CrossRef]

9. Wu, K.; Liu, G.Q.; Wang, L.; Xu, B.F. Research on new rapid and deep plasma nitriding techniques of AISI 420 martensitic stainless steel. *Vacuum* **2010**, *84*, 870–875. [CrossRef]

10. Cleugh, D.; Blawert, C.; Steinbach, J.; Ferkel, H.; Mordike, B.L.; Bell, T. Effects of rare earth additions on nitriding of EN40B by plasma immersion ion implantation. *Surf. Coat. Technol.* **2001**, *142–144*, 392–396. [CrossRef]

11. Medina, A.; Aguilar, C.; Béjar, L.; Oseguera, J.; Ruíz, A.; Huape, E. Effects of post-discharge nitriding on the structural and corrosion properties of 4140 alloyed steel. *Surf. Coat. Technol.* **2019**, *366*, 248–254. [CrossRef]

12. Dai, M.; Li, C.; Hu, J. The enhancement effect and kinetics of rare earth assisted salt bath nitriding. *J. Am. Ceram. Soc.* **2016**, *688*, 350–356. [CrossRef]

13. Cheng, X.-H.; Xie, C.-Y. Effect of rare earth elements on the erosion resistance of nitrided 40Cr steel. *Wear* **2003**, *254*, 415–420. [CrossRef]

14. Tang, L.N.; Yan, M.F. Effects of rare earths addition on the microstructure, wear and corrosion resistances of plasma nitrided 30CrMnSiA steel. *Surf. Coat. Technol.* **2012**, *206*, 2363–2370. [CrossRef]

15. Chen, X.; Xiangyun, B.; Xiao, Y.; Chengsong, Z.; Tang, L.; Cui, G.; Yang, Y. Low-temperature gas nitriding of AISI 4140 steel accelerated by $LaFeO_3$ perovskite oxide. *Appl. Surf. Sci.* **2019**, *466*, 989–999. [CrossRef]

16. Zhang, C.S.; Yan, M.F.; Sun, Z. Experimental and theoretical study on interaction between lanthanum and nitrogen during plasma rare earth nitriding. *Appl. Surf. Sci.* **2013**, *287*, 381–388. [CrossRef]

17. Liu, R.L.; Yan, M.F.; Wu, D.L. Microstructure and mechanical properties of 17-4PH steel plasma nitrocarburized with and without rare earths addition. *J. Mater. Res.* **2010**, *210*, 784–790. [CrossRef]

18. Liu, R.L.; Qiao, Y.J.; Yan, M.F.; Fu, Y.D. Layer growth kinetics and wear resistance of martensitic precipitation hardening stainless steel plasma nitrocarburized at 460A degrees C with rare earth addition. *Met. Mater. Int.* **2013**, *19*, 1151–1157. [CrossRef]

19. You, Y.; Yan, J.H.; Yan, M.F. Atomistic diffusion mechanism of rare earth carburizing/nitriding on iron-based alloy. *Appl. Surf. Sci.* **2019**, *484*, 710–715. [CrossRef]

20. Anjos, A.D.; Scheuer, C.J.; Brunatto, S.F.; Perito Cardoso, R. Low-temperature plasma nitrocarburizing of the AISI 420 martensitic stainless steel. Microstructure and process kinetics. *Surf. Coat. Technol.* **2015**, *275*, 51–57. [CrossRef]

21. Alphonsa, I.; Chainani, A.; Raole, P.M.; Ganguli, B.; John, P.I. A study of martensitic stainless steel AISI 420 modified using plasma nitriding. *Surf. Coat. Technol.* **2002**, *150*, 263–268. [CrossRef]

22. Liu, T.K.; Qian, J.N.; Yao, Y.Y.; Shi, Z.; Han, L.; Liang, C.; Li, B.; Dong, L.; Fan, M.; Zhang, L. Research on SCR of NO with CO over the $Cu_{0.1}La_{0.1}Ce_{0.8}O$ mixed-oxide catalysts: Effect of the grinding. *Mol. Catal.* **2017**, *430*, 43–53. [CrossRef]

23. Liu, R.L.; Yan, M.F. Effects of rare earths on nanocrystalline for nitrocarburised layer of stainless steel. *Mater. Struct.* **2017**, *33*, 1346–1351. [CrossRef]

Article

Acceleration of Plasma Nitriding at 550 °C with Rare Earth on the Surface of 38CrMoAl Steel

Dongjing Liu [1], Yuan You [1,2,*], Mufu Yan [2], Hongtao Chen [3], Rui Li [1], Lin Hong [1] and Tingjie Han [1]

1   School of Materials Science and Engineering, Qiqihar University, Qiqihar 161006, China;
     lovccat@163.com (D.L.); ly731374908@163.com (R.L.); hl25802021@163.com (L.H.); htj202021@163.com (T.H.)
2   National Key Laboratory of Metal Precision Thermal Processing, School of Materials Science and Engineering,
     Harbin Institute of Technology, Harbin 150001, China; yanmufu@hit.edu.cn
3   School of Materials Science and Engineering, Harbin University of Science and Technology,
     Harbin 150040, China; htchen83@163.com
*   Correspondence: youyuan@qqhru.edu.cn

Abstract: In order to explore the effect of the addition of rare earth (RE) to a steel microstructure and the consequent performance of a nitrided layer, plasma nitriding was carried out on 38CrMoAl steel in an atmosphere of NH_3 at 550 °C for 4, 8, and 12 h. The modified layers were characterized using an optical microscope (OM), a microhardness tester, X-ray diffraction (XRD), a scanning electron microscope (SEM), a transmission electron microscope (TEM), and an electrochemical workstation. After 12 h of nitriding without RE, the modified layer thickness was 355.90 µm, the weight gain was 3.75 mg/cm^2, and the surface hardness was 882.5 HV$_{0.05}$. After 12 h of RE nitriding, the thickness of the modified layer was 390.8 µm, the weight gain was 3.87 mg/cm^2, and the surface hardness was 1027 HV$_{0.05}$. Compared with nitriding without RE, the ε-Fe$_{2-3}$N diffraction peak was enhanced in the RE nitriding layer. After 12 h of RE nitriding, La, LaFeO$_3$, and a trace amount of Fe$_2$O$_3$ appeared. The corrosion rate of the modified layer was at its lowest (15.089×10^{-2} mm/a), as was the current density (1.282×10^{-5} A/cm^2); therefore, the corrosion resistance improved.

Keywords: 38CrMoAl steel; plasma nitriding; microhardness; corrosion resistance; modified layer

Citation: Liu, D.; You, Y.; Yan, M.; Chen, H.; Li, R.; Hong, L.; Han, T. Acceleration of Plasma Nitriding at 550 °C with Rare Earth on the Surface of 38CrMoAl Steel. *Coatings* **2021**, *11*, 1122. https://doi.org/10.3390/coatings11091122

Academic Editor: Kyong Yop Rhee

Received: 3 August 2021
Accepted: 10 September 2021
Published: 16 September 2021

Publisher's Note: MDPI stays neutral with regard to jurisdictional claims in published maps and institutional affiliations.

Copyright: © 2021 by the authors. Licensee MDPI, Basel, Switzerland. This article is an open access article distributed under the terms and conditions of the Creative Commons Attribution (CC BY) license (https://creativecommons.org/licenses/by/4.0/).

1. Introduction

Plasma nitriding is a widely used chemical heat treatment technology. This method uses the active nitrogen atoms generated during the cathode sputtering process at a relatively low temperature (350–570 °C) to accumulate on the surface of a workpiece, diffuse into the substrate, and finally form a nitriding-modified layer with excellent performance. For a variety of alloy materials, it can improve surface hardness, fatigue resistance, wear resistance, and corrosion resistance [1–7]. At present, the methods most commonly used to increase the thickness of the nitride layer are to increase the nitriding temperature or to extend the process time [8]. However, these methods have some disadvantages, such as the coarsening of the structure and reduced hardness [9,10]. Therefore, the question of how to increase the thickness of the modified layer in a short time presents a difficult challenge. Researchers have found that adding some catalysts can improve the efficiency of nitriding. Among them, rare earth (RE) elements have been proven to be effective catalysts for chemical heat treatment [11].

In the early 1980s, researchers studied the chemical heat treatment of RE. In the following decades, scholars explored the effects of RE catalysis and microalloying [12]. Many researchers observed that in the process of gas carburizing, gas carbonitriding, gas nitrocarburizing, and plasma nitriding, RE elements can diffuse deeper to the surface [13–15]. At the same time, the involvement of RE elements causes the nitrogen to diffuse deeper into the nitrided layer. Yan et al. [16,17] showed that rare earth elements induced nitriding and observed similar catalytic effects; they also studied the general catalytic effect of rare earth elements on the nitriding process of various steels, including stainless steel [18,19] and alloy steel [20–22].

Due to its good wear resistance, high fatigue strength and high strength, 38CrMoAl steel is widely used in many industries [23,24]. However, in some applications, it cannot meet the requirements of high surface hardness and corrosion resistance. Therefore, surface modification approaches are used to improve the structure and properties of 38CrMoAl steel. However, there are few studies on the surface modification of 38CrMoAl steel, especially with the addition of RE. Therefore, it is very important to study the effect of plasma nitriding following the addition of RE on the thickness, surface hardness, phase composition, and corrosion resistance of a modified 38CrMoAl steel layer.

2. Materials and Methods

The test material was 38CrMoAl steel with a thickness of 5 mm and a diameter of ϕ 20 mm, and its chemical compositions are shown in Table 1. The steel was solution-treated in an SXL-1400C box-type electric furnace and austenitized at 940 °C for 1 h, after which oil quenching treatment was performed. A QG-1 metallographic cutting machine was used to cut the solution-treated sample into 5 mm thin slices. The surface was polished with 120# water sandpaper. Next, absolute ethanol was used for ultrasonic cleaning for 10–15 min. The sample was then blow dried with a hair dryer before being wiped with a dust-free cloth, weighed, and tied together with a thin iron wire. The $1 \times 1 \times 1$ cm^3 rare earth (RE) lanthanum was cut into 1/8 cube blocks, and the RE was hung on the cathode platform with the iron wire, in order to maximize the sputtering of RE.

Table 1. Chemical composition of 38CrMoAl steel (wt.%).

Element	C	Si	Mn	Cr	Mo	Al	Fe
Content	0.38	0.31	0.45	1.67	0.20	0.88	Bal.

Before plasma nitriding in the LDMC-30AFZ ion nitriding furnace, the furnace was evacuated to below 10 pa. The nitriding atmosphere was NH_3 (100 mL/min) at 550 °C for 4, 8, and 12 h. A CPA-225D electronic balance (Sai Dolis scientific instruments Beijing co., Ltd., Beijing, China) was used to measure the mass before and after the nitriding sample. After nitriding, the surface of the sample was polished with 120#-2000# water sandpaper. The polished sample was corroded with 4% nitric acid alcohol solution for 15 s, rinsed with alcohol, and dried. The cross-section structure and thickness of the modified layer was observed with a 9XB-PC metallographic microscope, the surface morphology was observed with a ThermoScientific™Apreo C (TEM, JEOL Ltd, Tokyo, Japan) model field emission scanning electron microscope, and a HV-1000IS microhardness tester (Shanghai Jvjing Precision Instrument Manufacturing Co., Ltd., Shanghai, China) was used to test the hardness of the modified layer. The phase composition of the sample surface after nitrogen treatment was characterized by a D8 Advance X-ray diffractometer (Bruker, Karlsruhe, Germany). Anodic polarization tests were employed to estimate the corrosion resistance of the specimens in a 3.5 wt.% NaCl solution. The reference electrode was Ag/AgCl and a platinum column was used as the auxiliary electrode. A corrosion resistance test was performed in a CHI604E B16276 electrochemical workstation.

3. Results and Discussion

3.1. Microstructure and Phase Composition of Modified Layer

Figure 1 shows the metallographic microstructure of the modified layer of 38CrMoAl steel without rare earth (RE) and with rare earth. The nitrided layer without RE was clearly divided from the matrix, and the microstructure of the nitrided layer was uniform. After nitriding for 4, 8, and 12 h, the thickness of the modified layer was 222.2, 345.7, and 355.9 μm, respectively; the thickness of the modified layer increased gradually with the extension of nitride time. However, at the same temperature of 550 °C, after the addition of RE, the nitrided layer thickened with time to 252.4, 358.0, and 391.8 μm (4, 8, 12 h), respectively., The thickened histograms of the nitrided modified layer with RE and without RE after nitriding for 4, 8, and 12 h are shown in Figure 2. The modified layer was thickened by 30.2, 12.3, and 34.9 μm at 4, 8, and 12 h, respectively. Compared with the nitriding layer without the addition of RE, the thickening rates of the RE nitriding layer were 13.6%, 3.50% and 9.80%, respectively, indicating that it was easier to obtain a thicker nitriding layer with the addition of RE at the same nitriding time intervals. The addition of RE in nitrided steel increased the concentration of microscopic crystal defects, such as vacancies and dislocations, in the modified layer and caused lattice distortion of the iron atoms around it, which was beneficial for the adsorption and diffusion of the nitrogen atoms. Therefore, in a short period of time, a higher nitrogen concentration was formed in the matrix, providing a very high nitrogen potential and concentration gradient for the inward diffusion of nitrogen atoms, which also provided good thermodynamic conditions for the diffusion of nitrogen atoms.

As shown in the XRD patterns of Figure 1a–c, which correspond to Figure 3a without RE nitridating, the main phase of the modified layer without RE nitridation was γ'-Fe_4N and the secondary phase was ε-$Fe_{2-3}N$. However, La and $LaFeO_3$ oxides were detected in the XRD patterns shown in Figure 1d–f, corresponding to Figure 3b with the addition of RE nitridation for 12 h. Because $LaFeO_3$ showed a strong adsorption effect on N atoms [25], the thickening rate of the modified layer increased with the addition of RE nitriding within the same timeframe.

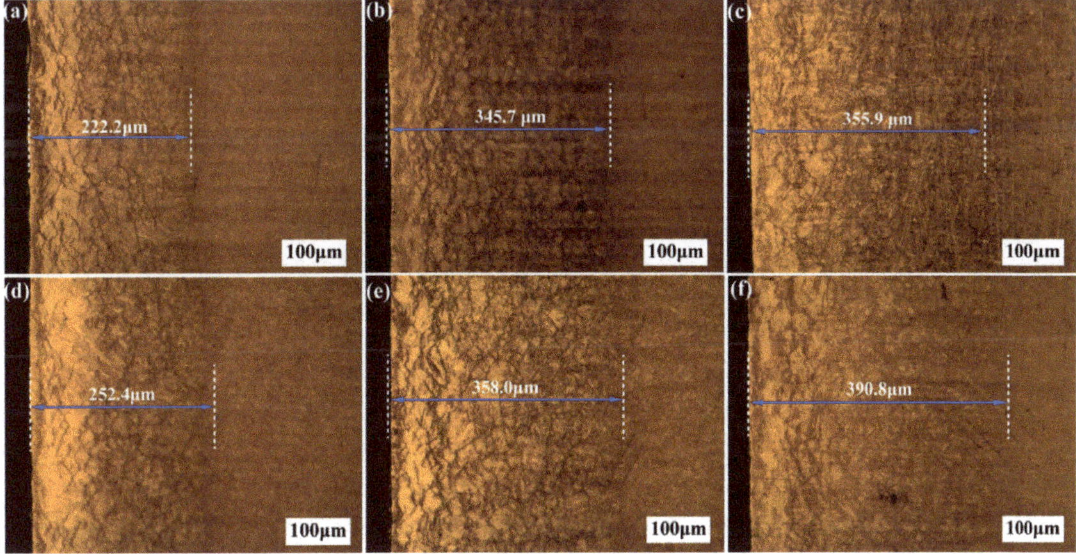

Figure 1. Microstructure of modified layer of 38CrMoAl steel under different nitriding conditions: (**a**) PN, 4 h; (**b**) PN, 8 h; (**c**) PN, 12 h; (**d**) RE, 4 h; (**e**) RE, 8 h; (**f**) RE, 12 h.

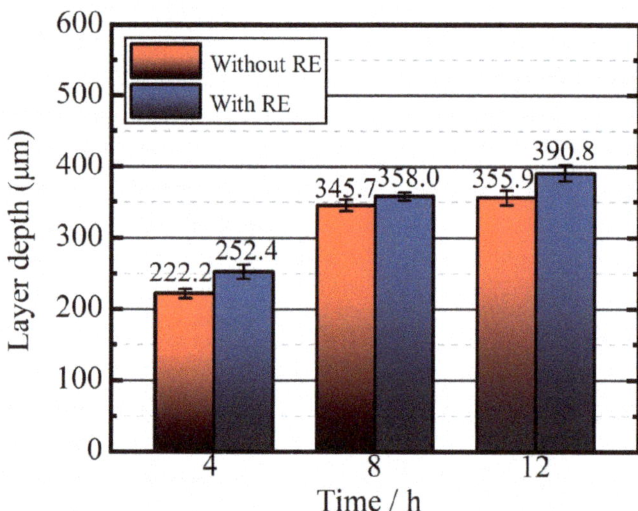

Figure 2. Cross-section thickening histogram of nitrided layer of 38CrMoAl Steel without and with RE.

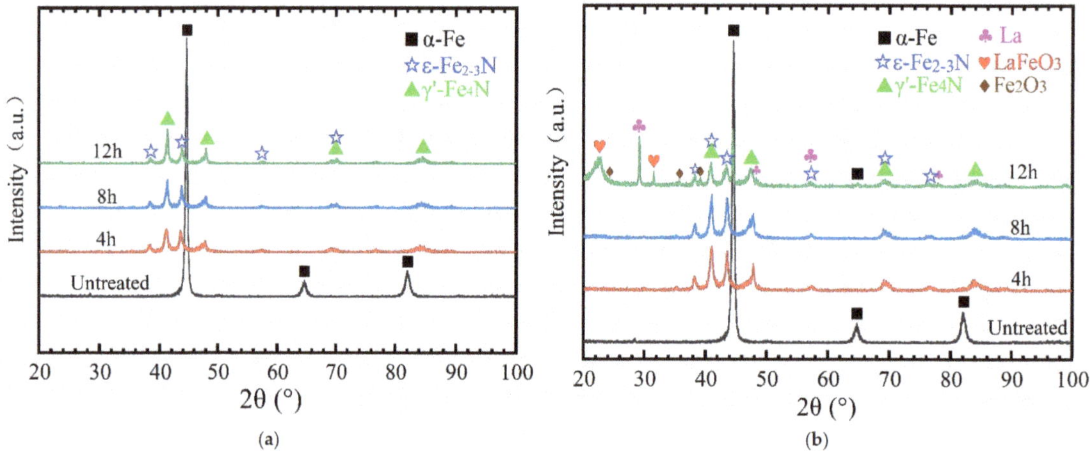

Figure 3. XRD pattern of 38CrMoAl steel without and with RE nitriding. (**a**) Without RE (**b**) With RE.

3.2. Weight Gain and Microhardness of Modified Layer

Figure 4 shows the histogram (4, 8, and 12 h) of the thickening of the steel without the RE-nitridated layer. The weight gain was 2.42, 3.35 and 3.75 mg/cm^2 at each respective time interval. Figure 5 shows the microhardness diagram of the modified layer. After 4 h of nitriding, the surface hardness was 701.3 HV$_{0.05}$. After nitriding for 8 h, the surface hardness was 975.3 HV$_{0.05}$. After 12 h of nitriding, the surface hardness of the modified layer was 882.5 HV$_{0.05}$.

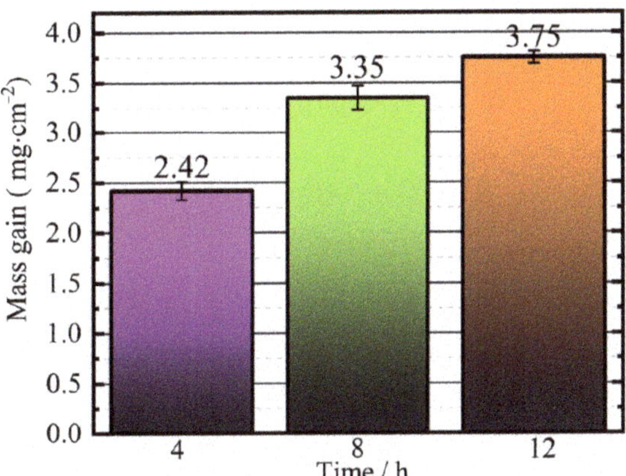

Figure 4. Histogram of weight gain of modified layer under nitriding conditions without RE.

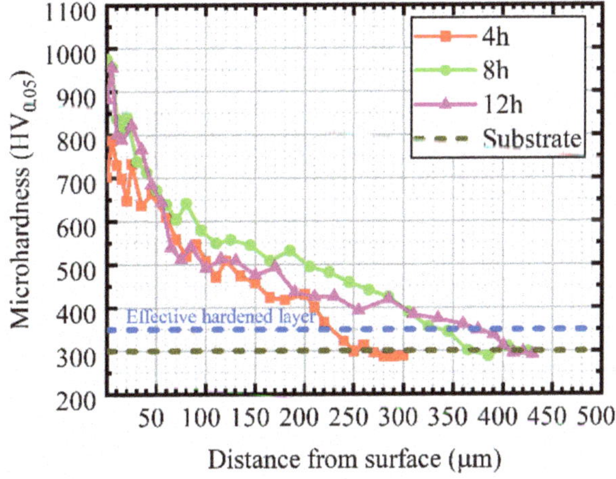

Figure 5. Microhardness of modified layer cross-section under nitriding conditions without RE.

Figure 6 shows the histogram of the weight gain after nitriding with the addition of for 4, 8, and 12 h. The weight gain was 2.49, 3.43, and 3.87 mg/cm^2, respectively. Figure 7 shows the microhardness distribution under nitriding conditions with RE. After nitriding for 4, 8, and 12 h, the surface hardness was 840.9, 989.6, and 1027.0 HV$_{0.05}$, respectively, and the surface hardness reached its peak at 12 h. According to the microhardness bar chart of the nitriding surface without and with rare earth in Figure 8, the increase percentage of the surface hardness after nitriding with RE over time was 19.7%, 1.47%, and 16.3%, for each respective time interval. Therefore, the surface hardness improved after the addition of RE nitriding. This was mainly because the addition of RE refines the microstructure of the nitrided layer and adjusts the phase proportion [26]. Moreover, there are dispersion strengthening and solid solution strengthening effects after adding rare earth nitriding. At the same time, the infiltration of rare earth elements leads to serious lattice distortion and increases dislocation around the boundary, resulting in grain boundary strengthening.

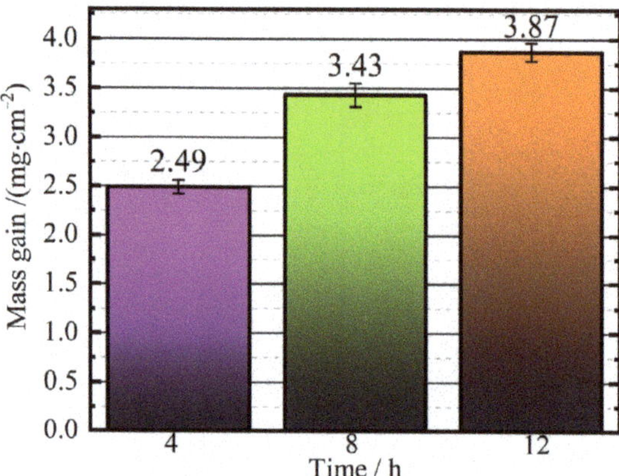

Figure 6. Bar graph of weight gain of modified layer under nitriding conditions with RE.

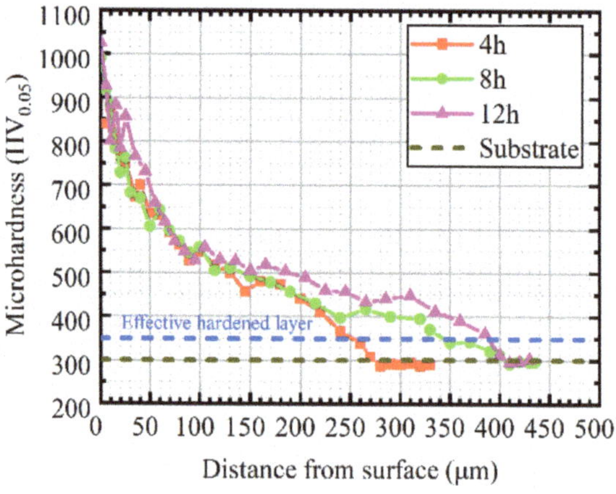

Figure 7. Microhardness of modified layer cross-section under nitriding conditions with RE.

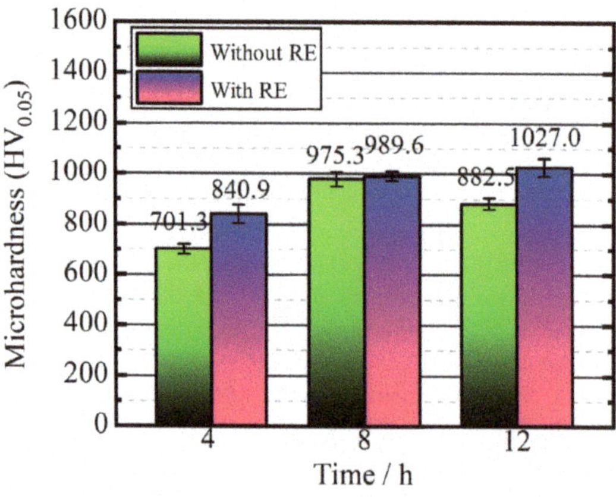

Figure 8. Histogram of the surface microhardness of the modified layer under nitriding conditions with and without RE.

3.3. Element Distribution in Modified Layer and EDS Energy Spectrum Analysis

The modified surface layers of the 12 h samples without RE nitriding were magnified by 100. Point scanning and plane scanning were performed for a, b, c, d, and e in Figure 9a by an EDS analyzer. Figure 9b1–b5 and Figure 9c1–c4 show the modified layer EDS spectrum test results. The distance from the surface of the modified layer was gradually transferred to the matrix, and the content distribution of each element was uneven. In combination with the line scan and plane scan in Figure 9a,b1–b5, it is also shown that under nitriding conditions without RE, the content of the element N in the modified layer didnot change significantly, and the distribution of the elements C, Fe, and Cr in the modified layer was relatively uniform.

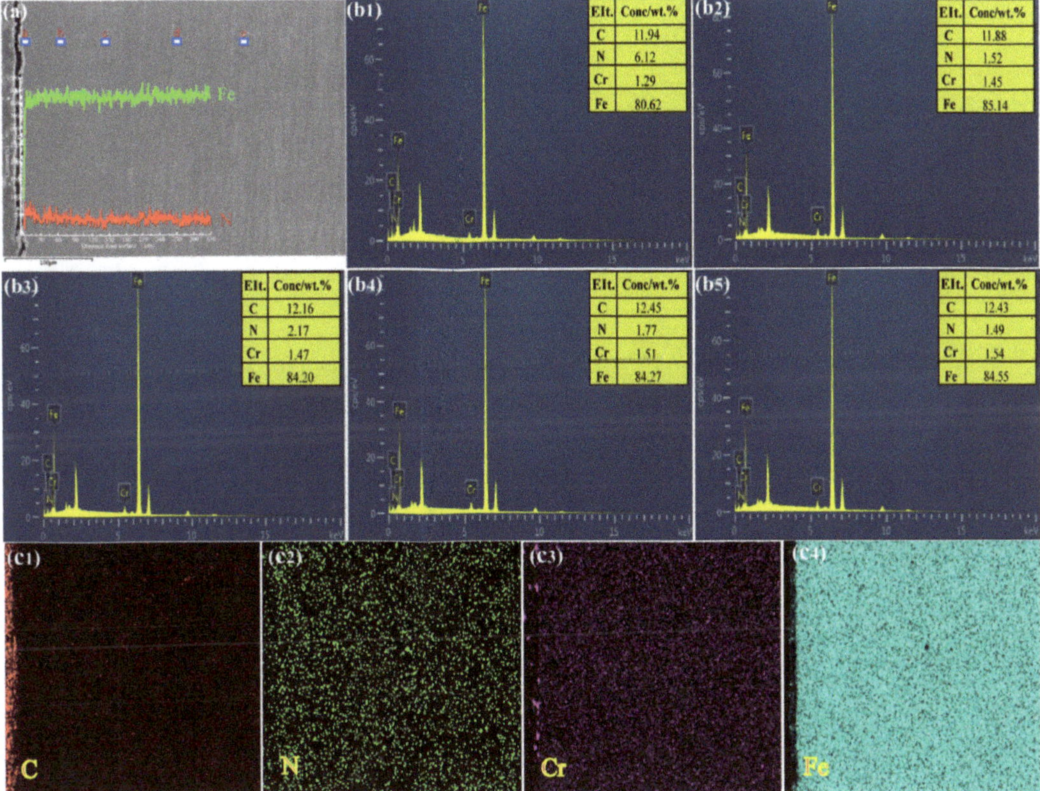

Figure 9. Element distribution diagram of the 12 h-nitrided layer without the addition of RE. (**a**) Fe, N, line-scan; (**b1–b5**) a,b,c,d,e points correspond to dot energy spectrum; (**c1–c4**) C, N, Cr, Fe element surface scanning.

Figure 10 shows the surface morphology and elements content after 12 h of RE nitriding. As can be seen from Figure 10b1–b5, the La content in the modified layer was small, while the N content showed a downward trend. At position c in Figure 10a, dot analysis was performed at 130.60 μm of the permeable layer, and the content of La was consistent with the surface content of the nitrided layer, indicating that RE atoms can be diffused into the modified layer. This was confirmed by mid-plane scanning, shown in Figure 10d5, but with the thickening of the nitrided layer, the modified layer contained a trace of La. Comparison between Figures 9b1 and 10b1 shows that the content of N element on the surface without RE nitriding was higher than it was after RE nitriding, although the thickness of the modified layer on the surface without RE nitriding was thinner than that of the layer subjected to nitriding with RE [27]. This indicates that the increase of the nitrogen concentration or nitrogen the potential did not necessarily lead to the thickening of the modified layer. The thickness of the modified layer is not only related to the external N potential, but also to the internal diffusion of N atoms [28].

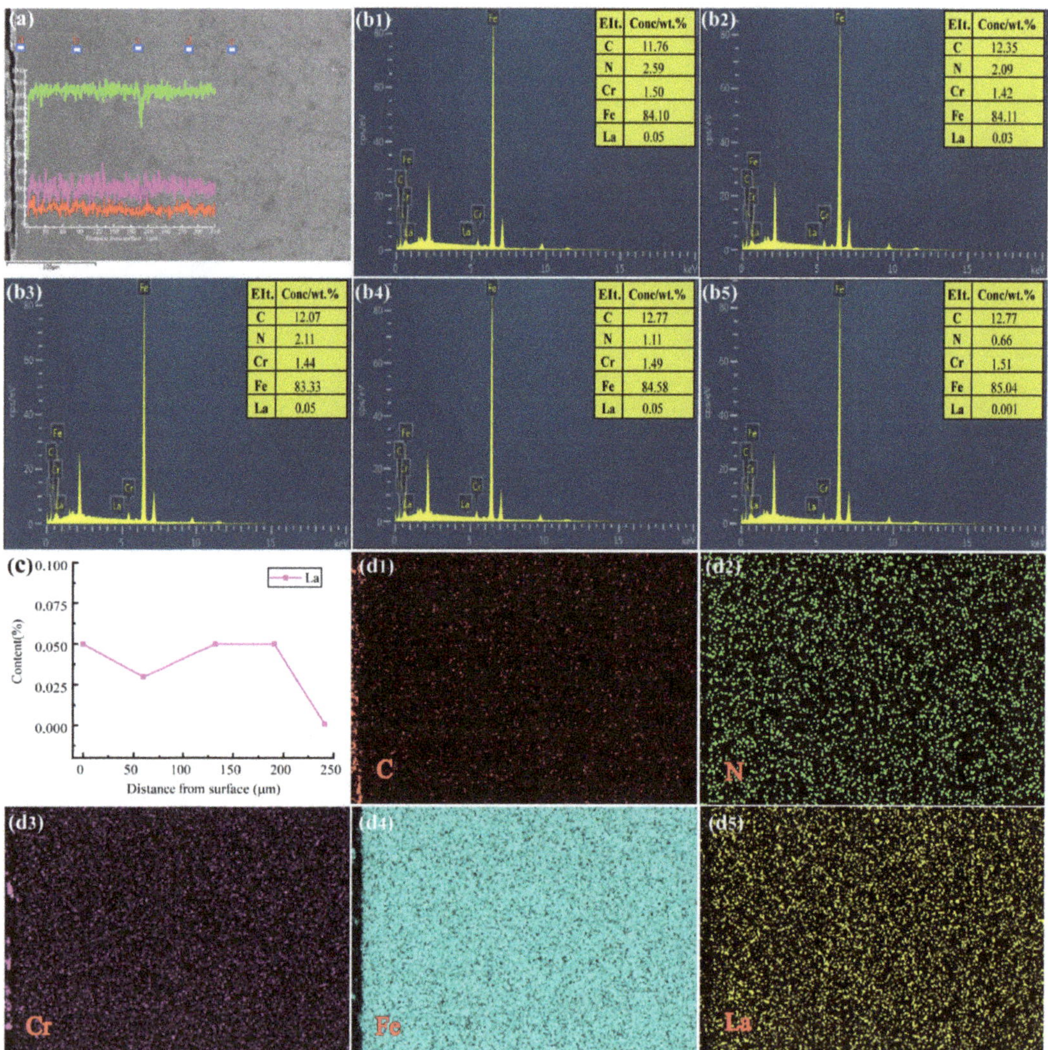

Figure 10. Element distribution diagram in 12 h-nitrided layer with the addition of RE. (**a**) Fe, N, line-scan; (**b1–b5**) a.b,c,d,e points correspond to dot energy spectrum; (**c**) The distance between La and the surface of the modified layer; (**d1–d5**) C, N, Cr, Fe, La element surface scanning.

The surface structure of the 12 h RE-nitrided samples was further studied by transmission electron microscopy (TEM). The bright-field image and the corresponding selected region electron diffraction (SAED) map are shown in Figure 11. A large number of nanocrystals were observed, as shown in Figure 11a. The size of the precipitated phase was extremely small, with the maximum size no greater than 60 nm. The results showed that the addition of RE refined the grain size and produced a microalloying effect on the modified layer. Figure 12b shows a clear diffraction ring caused by a large number of nanocrystals, indicating that these La, γ'-Fe$_4$N, ε-Fe$_{2\text{-}3}$N, and La phases existed in the outermost layer in the form of nanocrystals. The size of these phases was generally limited to 100 nm, so they can be referred to as nanophases. Yan et al. [29,30] obtained similar results with plasma nitriding, and found that the existence of the nano-phase helps to improve the hardness and wear resistance of the nitriding layer. Furthermore, these phases are consistent with the XRD results in Figure 3b.

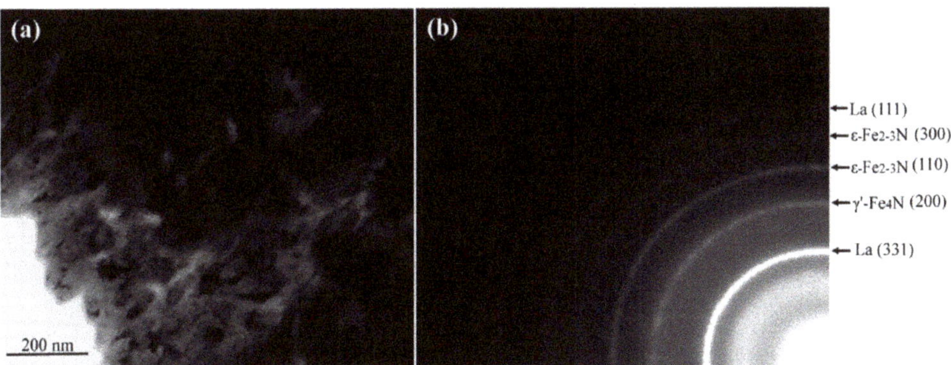

Figure 11. TEM image after rare earth nitriding for 12h. (**a**) Bright-field image; (**b**) Corresponding SAED.

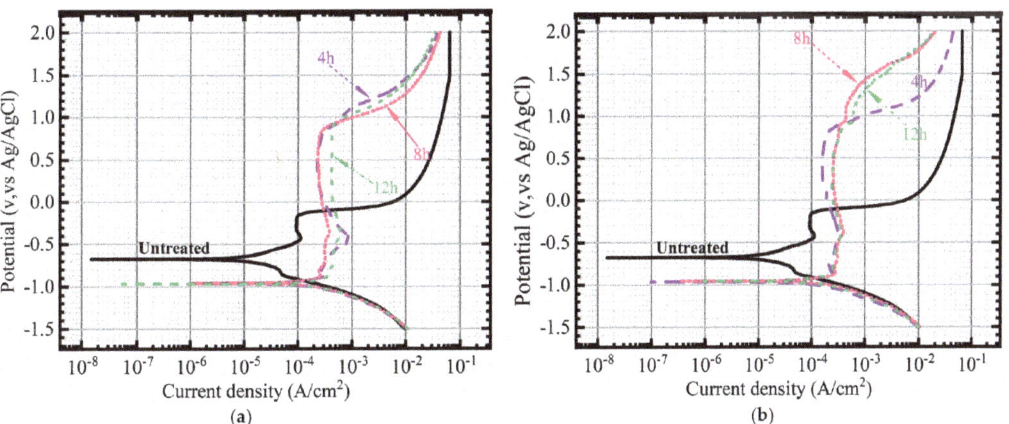

Figure 12. Corrosion resistance curve of modified layer under different nitriding conditions (4, 8, 12 h): (**a**) PN; (**b**) RE.

3.4. Polarization Curve and Fitting Data

The polarization curve data in Figure 12a were used for fitting, and the results are shown in Table 2. After nitriding for 4, 8, and 12 h, the corrosion rates of the modified layer were 21.405×10^{-2} mm/a, 21.939×10^{-2} mm/a, and 25.714×10^{-2} mm/a, respectively. The corrosion rates peaked after 12 h of nitriding. The current corrosion density was also increased to 2.186×10^{-5} A/cm^2 compared with that of the non-nitrided sample. After fitting the polarization curve data shown in Figure 12b, we obtained the fitting results shown in Table 3. After nitriding with RE for 4, 8, and 12 h, the corrosion rates of the modified layer were 16.092×10^{-2} mm/a, 20.754×10^{-2} mm/a, and 15.089×10^{-2} mm/a, respectively; the corrosion rate decreased, increased and then decreased again. The corrosion rates of the RE-nitrided samples reached their minimum at 12 h, and the current density of the RE-nitrided samples decreased to 1.282×10^{-5} A/cm^2 compared with the samples without RE, and the width of passivation zone was wider. Tang [25] discussed the main reasons for the enhanced corrosion resistance of ε-nitride compounds on the workpiece surface. The X-ray diffraction analysis in Figure 3b shows that the presence of La and oxide LaFeO$_3$ significantly improved the intensity of the ε-Fe$_{2-3}$N diffraction peak and increased the content of the ε-phase in the samples. Because ε-nitrides are the most important parameters controlling the corrosion resistance of nitrided samples, the corrosion resistance of our rare earth-nitrided samples was enhanced at 12 h.

Table 2. Fitting data of modified 38CrMoAl steel layer under nitriding conditions without RE.

Polarization Curve Fitting Data Unit	Untreated	4 h	8 h	12 h
Corrosion rate ($\times 10^{-2}$ mm/a)	17.472	21.405	21.939	25.714
Rp($\times 10^3$ Ω/cm^2)	1.756	1.434	1.400	1.193
Io($\times 10^{-5}$ A/cm^2)	1.486	1.820	1.865	2.186
Eo/V	-0.674	-0.960	-0.955	-0.967
Passivation zone width/V	0.715	2.219	2.267	2.395

Table 3. Fitting data of modified 38CrMoAl steel layer under nitriding conditions with RE.

Polarization Curve Fitting Data Unit	Untreated	4 h	8 h	12 h
Corrosion rate ($\times 10^{-2}$ mm/a)	17.472	16.092	20.754	15.089
Rp ($\times 10^3$ Ω/cm^2)	1.756	1.906	1.479	2.034
Io ($\times 10^{-5}$ A/cm^2)	1.486	1.368	1.765	1.282
Eo/V	-0.674	-0.953	-0.954	-0.962
Passivation zone width/V	0.715	2.955	2.934	2.458

3.5. Thermodynamics of the Modified Layer

Because the nitrided atmosphere was NH$_3$, the main reaction in the furnace was:

$$NH_3 \rightarrow 3/2H_2 + [N] \tag{1}$$

The decomposition of ammonia produced [N] atoms (Equation (1)), some of which penetrated into the surface of the workpiece, and the other part formed N$_2$, which became a stable gas.

The nitride produced by the decomposition of iron nitride provided the conditions for the formation of rare earth nitride [31].

$$2\varepsilon\text{-Fe}_2\text{N} \rightarrow \text{Fe}_4\text{N} + 2[N] \tag{2}$$

$$\gamma'\text{-Fe}_4\text{N} \rightarrow 4\alpha\text{-Fe} + [N] \tag{3}$$

The RE formed compounds with nitrogen (N), a gap element of chemical heat treatment, and the bond changed from metal to ionic. Rare earths also reacted with hydrogen (H) to form rare earth hydrides.

Since the hydrogenation of rare earth can occur above 300 °C:

$$La^{+3} + 2Cl^- + H_2 = LaH_{2\text{-}3} + 3/2Cl_2 \tag{4}$$

Trace RE hydrides floating in the furnace flow improved the adsorption capacity of the workpiece surface diffusion. When RE hydrides are attracted to the workpiece surface, the following rare earth nitriding reactions occur:

$$LaH_2 + [N] \rightarrow La[N] + H_2 \tag{5}$$

The thermodynamic driving force of the decomposition of ε-Fe$_2$N (Reaction 2) was greater than that of γ'-Fe$_4$N (Reaction 3). Therefore, it is speculated that γ'-Fe4N was still in the modified layer [32]. The calculated negative $\Delta G1_T{}^0$ shows that the reaction occurs spontaneously from left to right in the range of 575–700 °C, forming active compounds of rare earth and nitrogen.

4. Conclusions

We studied the effect of the addition of rare earth (RE) to a modified layer of 38CrMoAl steel by plasma nitriding at 550 °C for 4, 8, and 12 h, and in an HN$_3$ atmosphere. The following conclusions were drawn:

(1) For nitriding without the addition of RE for 4, 8, and 12 h, the thickness of the modified layers was 222.2, 345.7, 355.9 μm for each respective timeframe, the weight gains were 2.42, 3.35, and 3.75 mg/cm^2, respectively, and the maximum surface hardness was 882.5 HV$_{0.05}$, at 12 h. After nitriding with the addition of RE for 4, 8, and 12 h, the modified layers were thickened by 30.2, 12.3 and 34.9 μm, respectively, and the weights were 2.49, 3.43 and 3.87 mg/cm^2, respectively. The maximum surface hardness was 1027.0 HV$_{0.05}$, at 12 h.

(2) After 12 h of nitriding with RE, La, LaFeO$_3$, and trace Fe$_2$O$_3$ appeared in the modified layer. The diffraction peak of γ'-Fe$_4$N in the RE nitriding layer was weakened compared with the layer nitrided without RE, while the diffraction peak of ε-Fe$_{2.3}$N was enhanced. Meanwhile, La, γ'-Fe$_4$N, ε-Fe$_{2.3}$N, and La phases existed in the outermost layer in the form of nanocrystals.

(3) After nitriding with the addition of RE for 12 h, the corrosion rate of the modified layer decreased to a minimum of 15.089×10^{-2} mm/a, and the current density also decreased to a minimum of 1.282×10^{-5} A/cm^2; consequently, the corrosion resistance increased.

Author Contributions: Conceptualization, D.L. and Y.Y.; methodology, D.L. and Y.Y.; software, L.H.; validation, T.H.; formal analysis, Y.Y.; investigation, D.L., R.L. and T.H.; resources, Y.Y. and M.Y.; data curation, Y.Y., T.H. and D.L.; writing—original draft preparation, Y.Y.; writing—review and editing, D.L., Y.Y. and H.C.; visualization, L.H.; supervision, M.Y.; project administration, Y.Y. and M.Y.; funding acquisition, Y.Y. and M.Y. All authors have read and agreed to the published version of the manuscript.

Funding: Supported by the National Natural Science Foundation of China (51401113) and Natural Science Foundation of Heilongjiang Province of China (E2016069), Heilongjiang Postdoctoral Financial Assistance (LBH-Z16061), The Fundamental Research Funds in Heilongjiang Provincial Universities (135309504).

Institutional Review Board Statement: Not applicable.

Informed Consent Statement: Not applicable.

Data Availability Statement: Data is contained within the article.

Conflicts of Interest: The authors declare no conflict of interest.

References

1. Bell, T.; Sun, Y.; Liu, Z.R.; Yan, M.F. Rare earth surface engineering. *Heat Treat. Met.* **2000**, *27*, 12–13.
2. Tuckart, W.; Forlerer, E.; Iurman, L. Delayed cracking in plasma nitriding of AISI 420 stainless steel. *Surf. Coat. Technol.* **2007**, *202*, 199–202. [CrossRef]
3. Yan, H.; Zhao, L.; Chen, Z.; Hu, X.; Yan, Z. Investigation of the surface properties and wear properties of AISI H11 steel treated by auxiliary heating plasma nitriding. *Coatings* **2020**, *10*, 528. [CrossRef]
4. Espitia, L.A.; Varela, L.; Pinedo, C.E.; Tschiptschin, A.P. Cavitation erosion resistance of low temperature plasma nitrided martensitic stainless steel. *Wear* **2013**, *301*, 449–456. [CrossRef]
5. Yang, J.; Liu, Y.; Ye, Z.; Yang, D.; He, S.; Li, X. Grease-lubricated tribological behaviour of nitrided layer on 2Cr13 steel in vacuum. *Appl. Surf. Sci.* **2010**, *256*, 4072–4080. [CrossRef]
6. Olzon-Dionysio, M.; Campos, M.; Kapp, M.; de Souza, S.; de Souza, S.D. Influences of plasma nitriding edge effect on properties of 316L stainless steel. *Surf. Coat. Technol.* **2010**, *204*, 3623–3628. [CrossRef]

7. Ahangarani, S.; Mahboubi, F.; Sabour, A.R. Effects of various nitriding parameters on active screen plasma nitriding behavior of a low-alloy steel. *Vacuum* **2006**, *80*, 1032–1037. [CrossRef]

8. Aghajani, H.; Torshizi, M.; Soltanieh, M. A new model for growth mechanism of nitride layers in plasma nitriding of AISI H11 hot work tool steel. *Vacuum* **2017**, *141*, 97–102. [CrossRef]

9. Karakan, M.; Alsaran, A.; Çelik, A. Effect of process time on structural and tribolo-gical properties of ferritic plasma nitrocarburized AISI 4140 steel. *Mater. Des.* **2004**, *25*, 349–353. [CrossRef]

10. Fattah, M.; Mahboubi, F. Comparison of ferritic and austenitic plasma nitriding and nitrocarburizing behavior of AISI 4140 low alloy steel. *Mater. Des.* **2010**, *31*, 3915–3921. [CrossRef]

11. Medina, A.; Aguilar, C.; Béjar, L.; Oseguera, J.; Ruíz, A.; Huape, E. Effects of post-discharge nitriding on the structural and corrosion properties of 4140 alloyed steel. *Surf. Coat. Technol.* **2019**, *366*, 248–254. [CrossRef]

12. Yuan, Z.X.; Yu, Z.S.; Tan, P.; Song, S.H. Effect of rare earths on the carburization of steel, Mater. *Sci. Eng. A* **1999**, *267*, 162–166. [CrossRef]

13. Yan, M.F. Study on absorption and transport of carbon in steel during gas carburizing with rare-earth addition. *Mater. Chem. Phys.* **2001**, *70*, 242–244. [CrossRef]

14. Yan, M.F.; Pan, W.; Bell, T.; Liu, Z. The effect of rare earth catalyst on carburizing kinetics in a sealed quench furnace with endothermic atmosphere. *Appl. Surf. Sci.* **2001**, *173*, 91–94. [CrossRef]

15. Chen, X.; Bao, X.; Xiao, Y.; Zhang, C.; Tang, L.; Yao, L.; Cui, G.; Yang, Y. Low-temperature gas nitriding of AISI 4140 steel accelerated by LaFeO3 perovskite oxide. *Appl. Surf. Sci.* **2019**, *466*, 989–999. [CrossRef]

16. Liu, R.L.; Yan, M.F. Effects of rare earths on nanocrystalline for nitrocarburised layer of stainless steel. *Mater. Struct.* **2017**, *33*, 1346–1351. [CrossRef]

17. Yan, M.F.; Liu, R.L. Influence of process time on microstructure and properties of17-4PH steel plasma nitrocarburized with rare earths addition at low temperature. *Appl. Surf. Sci.* **2010**, *256*, 6065–6071. [CrossRef]

18. Yan, M.F.; Liu, R.L. Martensitic stainless steel modified by plasma nitrocarburizing at conventional temperature with and without rare earths addition. *Surf. Coat. Technol.* **2010**, *205*, 345–349. [CrossRef]

19. Wang, X.; Yan, M.; Liu, R.; Zhang, Y. Effect of rare earth addition on microstructure and corrosion behavior of plasma nitrocarburized M50NiL steel. *J. Rare Earths* **2016**, *34*, 1148–1155. [CrossRef]

20. Wu, Y.; Yan, M. Effects of lanthanum and cerium on low temperature plasma ni-trocarburizing of nanocrystallized 3J33 steel. *J. Rare Earths* **2011**, *29*, 383–387. [CrossRef]

21. Liu, R.L.; Yan, M.F. The microstructure and properties of 17-4PH martensitic precipitation hardening stainless steel modified by plasma nitrocarburizing. *Surf. Coat. Technol.* **2010**, *204*, 2251–2256. [CrossRef]

22. Tang, L.N.; Yan, M.F. Effects of rare earths addition on the microstructure, wear and corrosion resistances of plasma nitrided 30CrMnSiA steel. *Surf. Coat. Technol.* **2012**, *206*, 2363–2370. [CrossRef]

23. Tong, W.P.; Han, Z.; Wang, L.M.; Lu, J.; Lu, K. Low-temperature nitriding of 38CrMoAl steel with a nanostructured surface layer induced by surface mechanical attrition treatment. *Surf. Coat. Technol.* **2008**, *202*, 4957–4963. [CrossRef]

24. Chen, Y. Lower temperature plasma nitriding without white layer for 38CrMoAl hydraulic plunger. *Mech. Eng.* **2017**, *53*, 81. [CrossRef]

25. Zhang, C.S.; Yan, M.F.; Sun, Z. Experimental and theoretical study on interaction between lanthanum and N during plasma rare earth nitriding. *Appl. Surf. Sci.* **2013**, *287*, 381–388. [CrossRef]

26. Cleugh, D.; Blawert, C.; Steinbach, J.; Ferkel, H.; Mordike, B.L.; Bell, T. Effects of rareearth additions on nitriding of EN40B by plasma immersion ion implantation. *Surf. Coat. Technol.* **2001**, *142–144*, 392–396. [CrossRef]

27. Wang, E.; Yang, H.; Wang, L. The thicker compound layer formed by different NH_3-N_2mixtures for plasma nitriding AISI 5140 steel. *J. Alloys Compd.* **2017**, *725*, 1320–1323. [CrossRef]

28. Mittemeijer, E.J.; Somers, M.A.J. Thermodynamics, kinetics, and process control of nitriding. *Surf. Eng.* **1997**, *13*, 483–497. [CrossRef]

29. Yan, M.F.; Wu, Y.Q.; Liu, R.L.; Yang, M.; Tang, L.N. Microstructure and mechanical properties of the modified layer obtained by low temperature plasma nitriding of nanocrystallized 18Ni maraging steel. *Mater. Des.* **2013**, *47*, 575–580. [CrossRef]

30. Yao, J.W.; Yan, F.Y.; Yan, M.F.; Zhang, Y.X.; Huang, D.M.; Xu, Y.M. The mechanism of surface nanocrystallization during plasma nitriding. *Appl. Surf. Sci.* **2019**, *488*, 462–467. [CrossRef]

31. Khalaj, G.; Nazari, A.; Khoie, S.; Khalaj, M.J.; Pouraliakbar, H. Chromium carbonitride coating produced on DIN 1.2210 steel by thermo-reactive deposition technique. *Surf. Coat. Technol.* **2013**, *225*, 1–10. [CrossRef]

32. Barin, I. *Thermochemical Data of Pure Substances*, 3rd ed; VCH Publishers: Weinheim, NY, USA, 1995.

 *coatings*

Article

The Effects of Transition Metal Oxides (Me = Ti, Zr, Nb, and Ta) on the Mechanical Properties and Interfaces of B₄C Ceramics Fabricated via Pressureless Sintering

Guanqi Liu [1], Shixing Chen [2], Yanwei Zhao [3], Yudong Fu [1,*] and Yujin Wang [2]

[1]   College of Materials Science and Chemical Engineering, Harbin Engineering University, Harbin 150001, China; liuguanqi@hrbeu.edu.cn
[2]   School of Materials Science and Engineering, Harbin Institute of Technology, Harbin 150001, China; 19S009067@stu.hit.edu.cn (S.C.); wangyuj@hit.edu.cn (Y.W.)
[3]   Science and Technology on Advanced Functional Composite Laboratory, Aerospace Research Institute of Materials & Processing Technology, Beijing 100076, China; ywzh227@163.com
*   Correspondence: fuyudong@hrbeu.edu.cn

Received: 10 November 2020; Accepted: 16 December 2020; Published: 18 December 2020

Abstract: There is little available research on how different transition metal oxides influence the behavior of B₄C-based ceramics, especially for Ta_2O_5 and Nb_2O_5. B_4C-MeB_2 (Me = Ti, Zr, Nb, and Ta) multiphase ceramic samples were prepared via in situ pressureless sintering at 2250 °C, involving the mixing of B_4C and MeO_x powders, namely TiO_2, ZrO_2, Nb_2O_5, and Ta_2O_5. The phase constituents, microstructures, and mechanical properties of the samples were tested. The results indicated that different transition metal elements had different effects on the ceramic matrix, as verified through a comparative analysis. Additionally, the doped WC impurity during the ball milling process led to the production of $(Me, W)B_2$ and W_2B_5, which brought about changes in morphology and performance. In this study, the Ta_2O_5-added sample exhibited the best performance, with elastic modulus, flexural strength, Vickers hardness, and fracture toughness values of 312.0 GPa, 16.3 GPa, 313.0 MPa, and 6.08 MPa·m$^{1/2}$, respectively. The comprehensive mechanical properties were better than the reported values when the mass fraction of the second phase was around five percent.

Keywords: pressureless sintering; boron carbide; transition metal oxide; multiphase ceramics

1. Introduction

Boron carbide (B_4C) is an essential structural ceramic due to its atomic structure. According to previous studies, the most notable properties of pure B_4C ceramic are its high melting point (2447 °C), extreme hardness (50 GPa), and low density (2.52 g/cm³) [1–3]. Although it exhibits attractive performance in various applications, the practical application of B_4C has been severely restricted as a result of its low fracture toughness and poor sinterability. These disadvantages are caused by the low self-diffusion coefficient and the dominance of covalent bonds in the B_4C atomic structure [4–8]. To solve this problem in order to improve the mechanical properties of multiphase ceramics, scholars and engineers have added various second-phase constituents [9–11].

According to previous research, the addition of IVB and VB transition metals, such as TiB_2, ZrB_2, NbB_2, and TaB_2, into the B_4C matrix can lead to densification and mechanical property improvements. These kinds of multiphase ceramics could be used as the initial bases for cutting tools, ballistic armor, thermal protection components, wear-resistant parts, and turbojet blades [12–15].

The conventional toughening methods used for B_4C typically involve laminated composite toughening, particle toughening, and whisker toughening [16,17]. Through whisker bridging, the whisker "pull-out" effect, and crack deformation, energy can be effectively consumed, thwarting crack propagation and improving the fracture toughness. For example, by adding 10 wt.% TiO_2 into B_4C at 1950 °C and 30 MPa, Wang et al. synthesized B_4C-TiB_2 composite ceramics with a relative density, flexural strength, and fracture toughness of 97.6%, 408.0 MPa, and 5.3 MPa·m$^{1/2}$, respectively [18]. Tamari et al. prepared composite ceramics of B_4C containing up to 30 vol% SiC whiskers by hot pressing at 2000–2200 °C under 30 MPa for 30 min. The Vickers hardness and elastic modulus were 30 and 430 GPa, respectively [19]. Jiang et al. prepared B_4C/40%BN laminated ceramic composites via the hot pressing process at 1850 °C for 1 h under 30 MPa pressure, for which the flexural strength was 245 MPa and the fracture toughness was 3.52 MPa·m$^{1/2}$ [20].

Compared with borides and carbides, oxides generate gas through in situ reactions with the matrix during the sintering process, meaning a uniform and refined structure can be obtained, resulting in materials with improved mechanical properties [21–25]. Although the addition of Ti, Zr, and other metal oxides to boron carbide has been extensively studied, transition metal oxides, including Ta and Nb, have not attracted much attention [26]. Moreover, no studies have been conducted that compare the effects of different transition metal oxides on boron carbide ceramics. Therefore, there is still space for further research in this direction.

Regarding the currently used sintering processes, the main methods include the hot press sintering method, spark plasma sintering (SPS) method, and pressureless sintering method [27–31]. Generally speaking, B_4C-TiB_2 composite ceramics possess a density of 98% when the temperature and pressure involved in the hot press sintering exceed 1957 °C and 30 MPa, respectively [32]. As for the spark plasma sintering method, a temperature of 1760 °C and pressure of 40 MPa are required to prepare materials of the same density [33]. Additionally, the preparation of B_4C-TiB_2 multiphase ceramics with pressureless sintering requires a temperature of more than 2150 °C [34]. Among the three methods, the hot press sintering method and spark plasma sintering method have stricter industrial production conditions, a more complicated process, and require more expensive equipment, with the required use of a mold being one of the other limitations. At the same time, hot press sintering synchronously involves more energy expenditure than the other two methods and results in increased inefficiency in industry. However, the cost of pressureless sintering is low and the operation method is simple, meaning products with complex shapes can be prepared, making this method suitable for mass production.

This study prepared B_4C-MeB_2 multiphase ceramics (Me = Ti, Zr, Nb, or Ta). For practical purposes, we chose to use the pressureless sintering method. We tested the mechanical properties and microstructures to determine how different transition metal oxides influence the behavior of B_4C-MeB_2 multiphase ceramics. The main aim of this project is to complement the existing research studies on B_4C-MeB_2 multiphase ceramics, with the results possibly providing useful references for further study.

2. Materials and Methods

2.1. Materials and Preparation

Table 1 shows the compositions of the samples used in this research. All of the B_4C-MeB_2 multiphase ceramics were fabricated by pressureless sintering at 2250 °C for 60 min under an Ar atmosphere, with a heating rate of 10 °C/min. The raw B_4C powder (Mudanjiang Diamond Boron Carbide Co., Ltd., Mudanjiang, China) used in this research had a purity of 98.5% and particle size of approximately 2.2~5 µm. The raw TiO_2, ZrO_2 Nb_2O_5, and Ta_2O_5 powders (Changsha Weihui High-Tech New Materials Co., Ltd., Changsha, China) had purities of 98.5% and particle sizes of approximately 2 µm. The B_4C-MeB_2 ceramics were comprised of 95 wt.% B_4C+5 wt.% MeO_x (Me = Ti, Zr, Nb, and Ta). After high-energy ball milling (Pulverisete 4) at a rate of 200 r/min with a grinding media/material ratio of 10:1, the size of the mixed powders was less than 1 µm on average. The mixed powders were dried and sieved through a 100-mesh screen in a flowing Ar atmosphere. Phenolic resin was added as molding binder.

Table 1. The starting compositions (in wt.%) and processing conditions of sintered compounds.

Sample	Composition (wt.%)		Preparation Conditions	
	B_4C	MeO_x	Temperature (°C)	Time (h)
B_4C	100	-		
$B_4C + TiO_2$	95	5		
$B_4C + ZrO_2$	95	5	2250	1
$B_4C + Nb_2O_5$	95	5		
$B_4C + Ta_2O_5$	95	5		

2.2. Experiments and Characterization

The Archimedes principle was used to measure the relative densities of the final samples. The phase constituents of the synthesized products were analyzed using X-ray diffraction (X'PERT, Panalytical, Almelo, The Netherlands) with Cu Kα radiation. The surface structures and fracture surfaces of products were observed with a scanning electron microscope (SEM, Merlin Compact, Carl Zeiss, Rauenstein, Germany). The compositions were observed using an energy-dispersive X-ray spectroscopy system (Helios NanoLab, FEI, Hillsboro, OR, USA). Microhardness values were determined using a Vickers indentation tester (HVS-1000Z, Shanghai, China) with a diamond indenter load of 9.8 N for 15 s. Flexural strength was measured using the three-point flexural method (3 mm × 4 mm × 36 mm) with a span of 30 mm and a crosshead speed of 0.5 mm/min. The fracture toughness was measured using the single-edge notched beam method (2 mm × 4 mm × 22 mm) across a span of 16 mm and with a crosshead speed of 0.05 mm/min.

3. Results and Discussion

3.1. Thermodynamic Calculations

For the reactions in this study, the Gibbs free energy values were calculated using FactSage software (version 8.0). B_2O_3 usually bonds to the surfaces of B_4C particles, reacting with B_4C during the heating process and releasing gas. The fly-off from the gas inhibits densification. The reaction also causes grain growth during ceramic sintering. The reactions between B_4C and MeO_x involved in the sintering process are as follows:

$$B_4C + 5B_2O_3 \rightarrow 7B_2O_2 + CO\uparrow \tag{1}$$

$$B_4C + 2TiO_2 + 3C \rightarrow 2TiB_2 + 4CO\uparrow \tag{2}$$

$$2ZrO_2 + B_4C + 3C \rightarrow 2ZrB_2 + 4CO\uparrow \tag{3}$$

$$Nb_2O_5 + B_4C + 4C \rightarrow 2NbB_2 + 5CO\uparrow \tag{4}$$

$$Ta_2O_5 + B_4C + 4C \rightarrow 2TaB_2 + 5CO\uparrow \tag{5}$$

$$C + O_2 \rightarrow 2CO\uparrow \tag{6}$$

$$C + 2B_2O_3 \rightarrow B_4C + 6CO\uparrow \tag{7}$$

$$5B_4C + 8WC \rightarrow 4W_2B_5 + 13C \tag{8}$$

We set up all of the reactions in order to find out how the phases change during the sintering process. Reactions (1)–(4) occur between B_4C and MeO_x (Me = Ti, Zr, Nb, and Ta), showing that all of the additives may cause in situ reactions. Reactions (5)~(6) are the reactions between the phenolic resin, as a carbon source, and residual B_2O_3 in the B_4C powder. The sintering densification was promoted due to the consumption of oxide on the surface of the B_4C, which inhibited grain growth. Figure 1 displays the Gibbs free energy values from 500 to 2500 °C for all of the reactions, which were calculated using FactSage 8.0. These results were in accordance with thermodynamics theories—all of the reactions occurred in this process. The final products mainly included B_4C and MeB_2 phases (Me = Ti, Zr, Nb, and Ta).

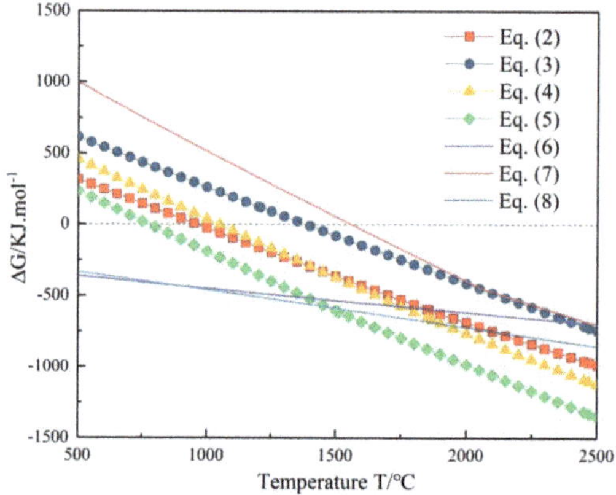

Figure 1. Gibbs free energy changes for each reaction as a function of temperature.

3.2. Phase Analysis

Figure 2 contains the XRD patterns of the final products, showing that the final samples were mainly composed of B_4C and MeB_2. Meanwhile, some graphite was retained as a residue. W_2B_5 was also present in the ZrO_2-added sample. The use of WC (Tungsten carbide) balls introduced some WC into the powder samples, which then reacted during sintering according to (7). The speculated WC was not found in the XRD patterns of samples, except in the ZrO_2-added sample. Further analysis of the energy spectrum was needed. Compared with diffraction standard cards (shown as red lines), it was found that the diffraction peaks of the ZrO_2-added, Nb_2O_5-added, and Ta_2O_5-added samples were shifted to higher angles, which may have been caused by W atoms migrating to the additive lattices to form $(Me, W)B_2$ solid solutions. Meanwhile, there were few transition metal oxides that were observed in the final products. The main phases of the final products were B_4C, $(Me, W)B_2$, and graphite. This result indicates that B_4C-MeB_2 multiphase ceramics were successfully fabricated in this research.

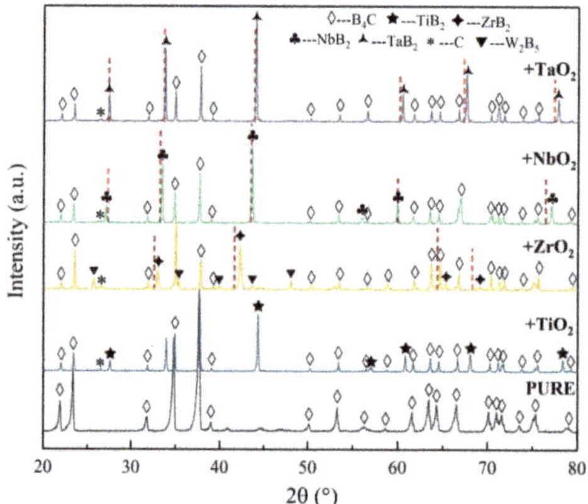

Figure 2. XRD pattern analysis for different sintered samples.

Table 2 shows the relative densities of the final products. The relative density of the pure sample was 78.3%. It can be clearly seen from the data in Table 1 that the addition of MeO_x significantly increased the samples' densification. Except for the TiO_2-added sample (relative density of 89.3%), the relative densities of other B_4C-MeB_2 multiphase ceramics exceeded 93%.

Table 2. The relative densities of the different samples.

No.	Sample Name	Relative Density (%)
1	B_4C	78.3
2	$B_4C+5\%TiO_2$	89.3
3	$B_4C+5\%ZrO_2$	93.2
4	$B_4C+5\%Nb_2O_5$	93.8
5	$B_4C+5\%Ta_2O_5$	94.0

Figure 3 shows the microsurfaces of different samples through SEM images. Overall, the samples contained several pores. The main component of the second phase was the $(Me, W)B_2$ solid solution. The resulting XRD patterns (Figure 2) and energy spectra (Figure 3) confirmed this. As shown in Figure 3a, the TiO_2-added sample consisted of a dark grey matrix, as attributed to B_4C, with the light grey second phase as attributed to $(Ti, W)B_2$. These second-phase particles were uniformly distributed but the number of pores was larger than in other particles. Additionally, the particle diameters for $(Ti, W)B_2$ samples were smaller than those of other samples. Figure 3b shows that there is a small number of pores in the ZrO_2-added sample, in which the light grey $(Zr, W)B_2$ solid-solution phase and bright white W_2B_5 phase were evenly distributed. Figure 3c shows that the Nb_2O_5-added sample exhibited a high density without distinct pores, in which the $(Nb, W)B_2$ solid-solution phase can be observed as a white phase. Figure 3d shows the Ta_2O_5-added sample, whose $(Ta, W)B_2$ solid-solution second phase had a similar distribution to the $(Nb, W)B_2$ solid-solution phase in the Nb_2O_5-added sample.

3.3. Mechanical Properties

Table 3 shows the elastic modulus and Vickers hardness values for different samples compared with the reported values. Firstly, the TiO_2-added and ZrO_2-added final products performed better, showing high elastic modulus values (>367 GPa) and hardness values (>19 GPa). Importantly, despite its lower relative density, the hardness of the TiO_2-added sample was only lower than that of the ZrO_2-added sample. Combined with the SEM analysis, it can be seen that the second-phase grains of the TiO_2-added sample were well distributed in the B_4C matrix. Therefore, although its density is low, the hardness of the TiO_2-added sample is still considerable due to the effect of fine-grain strengthening. For the ZrO_2-added sample, according to the shift in the XRD peaks, it can be seen that W is highly soluble in ZrB_2. Therefore, a $(Zr, W) B_2$ solid-solution phase was formed with a large W content. Due to the effect of solid-solution strengthening, the hardness of this sample was significantly improved, having the highest hardness value out of all samples.

Compared with the reference values in other reports, the hardness values of the samples in this experiment were similar to those obtained using SPS and hot press sintering and higher than those obtained using the pressureless method. This is related to the sintering method used and the form of the added compound. Although the temperature is the same as that used in other sintering methods, pressureless sintering lacks the energy provided by external mechanical pressure or electric currents, which may reduce the degree of densification. In addition, the in situ oxide reaction process releases gas, which eventually leads to more pores being formed in the material. Therefore, the hardness of the samples prepared in this study was expected to be slightly lower than the reference values. However, owing to the variable solubility and theoretical hardness of the second phase, the values obtained in this study were higher than expected.

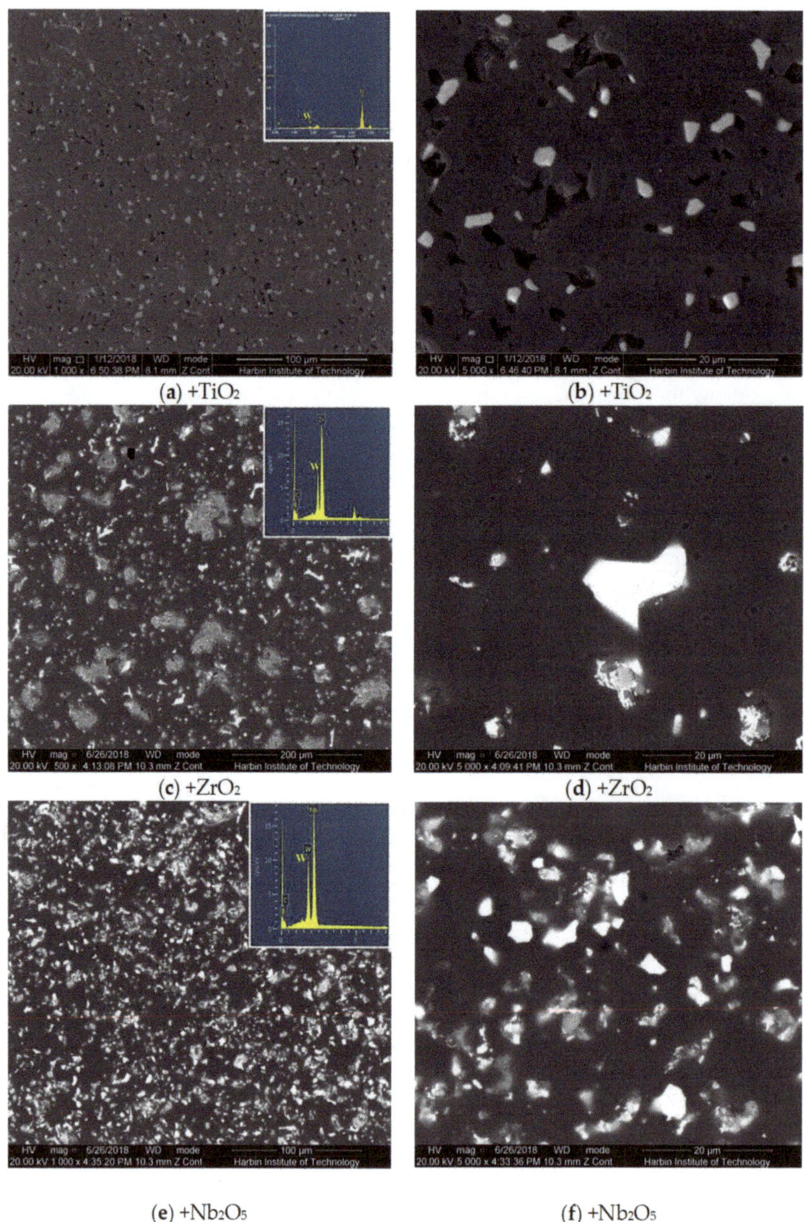

(**a**) +TiO$_2$ (**b**) +TiO$_2$

(**c**) +ZrO$_2$ (**d**) +ZrO$_2$

(**e**) +Nb$_2$O$_5$ (**f**) +Nb$_2$O$_5$

Figure 3. *Cont.*

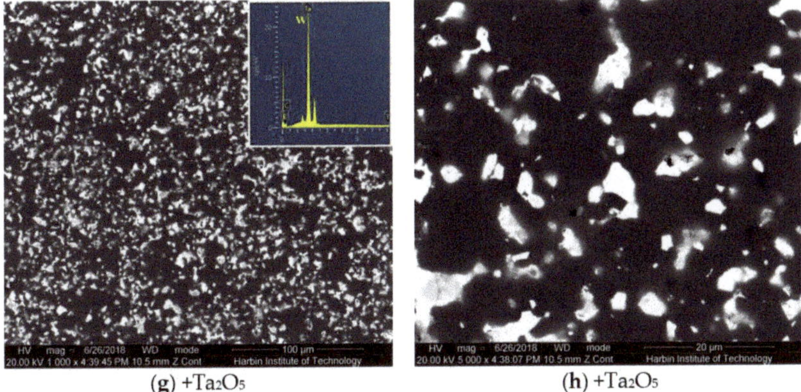

(g) +Ta$_2$O$_5$ (h) +Ta$_2$O$_5$

Figure 3. The magnification images of SEM analysis of B$_4$C-MeB$_2$ composites prepared from different additives (Me = Ti, Zr, Nb, or Ta): (**a,b**) +TiO$_2$ with different magnifications; (**c,d**) +ZrO$_2$ with different magnifications; (**e,f**) +Nb$_2$O$_5$ with different magnifications; (**g,h**) +Ta$_2$O$_5$ with different magnifications.

Table 3. Elastic modulus and Vickers hardness values for different samples as compared with the reported values [35–37].

No.	Process Condition		Sample Name	Elastic Modulus (GPa)	Hardness (GPa)
1			B$_4$C	209	12.4 ± 0.34
2			B$_4$C + 5%TiO$_2$	411	19.2 ± 3.34
3	In this study	2250 °C, 1 h	B$_4$C + 5%ZrO$_2$	367	21.1 ± 0.98
4			B$_4$C + 5%Nb$_2$O$_5$	296	15.0 ± 2.27
5			B$_4$C + 5%Ta$_2$O$_5$	312	16.3 ± 1.02
6	Xu et al. [35]	SPS, 1800 °C, 5 min, 50 MPa	B$_4$C + 2.8%TiB$_2$	-	17
7	Dudina et al. [36]	SPS, 1700 °C, 2 min, 100 MPa	B$_4$C + 23%Ti	-	19.5
9	Liu et al. [37]	2150 °C, 1 h	B$_4$C + 5%TiB$_2$	-	17

Table 4 shows the flexural strength and fracture toughness of B$_4$C ceramics prepared in the study.

Table 4. Flexural strength and fracture toughness values for the different samples and the reported values [38–40].

No.	Process Condition		Samples Composition	Flexural Strength (MPa)	Fracture Toughness (MPa·m$^{1/2}$)
1			B$_4$C	188 ± 4.38	1.98 ± 0.31
2			B$_4$C + 5%TiO$_2$	336 ± 21.7	3.75 ± 0.30
3	In this study	2250 °C, 1 h	B$_4$C + 5%ZrO$_2$	367 ± 24.9	4.06 ± 0.16
4			B$_4$C + 5%Nb$_2$O$_5$	268 ± 15.3	5.56 ± 0.38
5			B$_4$C + 5%Ta$_2$O$_5$	313 ± 11.7	6.08 ± 0.08
6	Wang et al. [18]	1850 °C, 1 h, 30 MPa	B$_4$C + 10%TiO$_2$	260	3.3
7	Demirskyi et al. [38]	SPS, 1800 °C, 1 min, 2350 °C, 1 min, 20 MPa	B$_4$C + 33%TaB$_2$	430	4.5
8	Liu et al. [39]	1600 °C, 1 h, 2060 °C, 0.5 h	B$_4$C + 16%ZrB$_2$	320	3.1
9	Ho et al. [40]	2150 °C, 1 h	B$_4$C + 5%TiB$_2$	260	2.6

The TiO$_2$-added and ZrO$_2$-added samples performed better in terms of flexural strength than the Nb$_2$O$_5$-added and Ta$_2$O$_5$-added samples, which was consistent with the elastic modulus and Vickers hardness. The fine crystals and solid solution led to the microstructure formation, which improved flexural strength. Owing to the lower relative density, the flexural strength of the TiO$_2$-added sample did not reach that of the ZrO$_2$-added sample.

In order to analyze the fracture mechanism of the material, we observed the microstructure morphology of the fractures through SEM images (as shown in Figure 4). The size and distribution of the second phase and the pores affected the mechanical properties of the samples. As shown in Figure 4a, the pores of the TiO_2-added samples had a uniform distribution. The second phase had an intergranular fracture mode and crystal grains were pulled out of the B_4C matrix. In the ZrO_2-added sample, the transgranular fracture in the B_4C matrix was effectively blocked by the ZrB_2 second-phase particles and pores. The transgranular fractures showed typical features, namely a river-like pattern, which can be clearly observed in Figure 4b (marked in yellow). The aggregation phenomenon was not observed in the distribution of pores. The pores in this sample were very small. Compared with large and aggregated pores, small pores led to a greater increase in the strength of the material. This was confirmed by the high flexural strength of the ZrO_2-added sample, as shown in Table 4. The fracture morphologies of the Nb_2O_5-added and Ta_2O_5-added samples are very similar. Compared with the former two samples, their B_4C matrix is denser. It is difficult to observe small and diffuse pores in the matrix (as shown in Figure 4c,d). Through analysis of the distribution of the second phase, it can be known that the second-phase particles had a "pinning effect" during the material fracture process, which is an excellent way to improve the flexural strength.

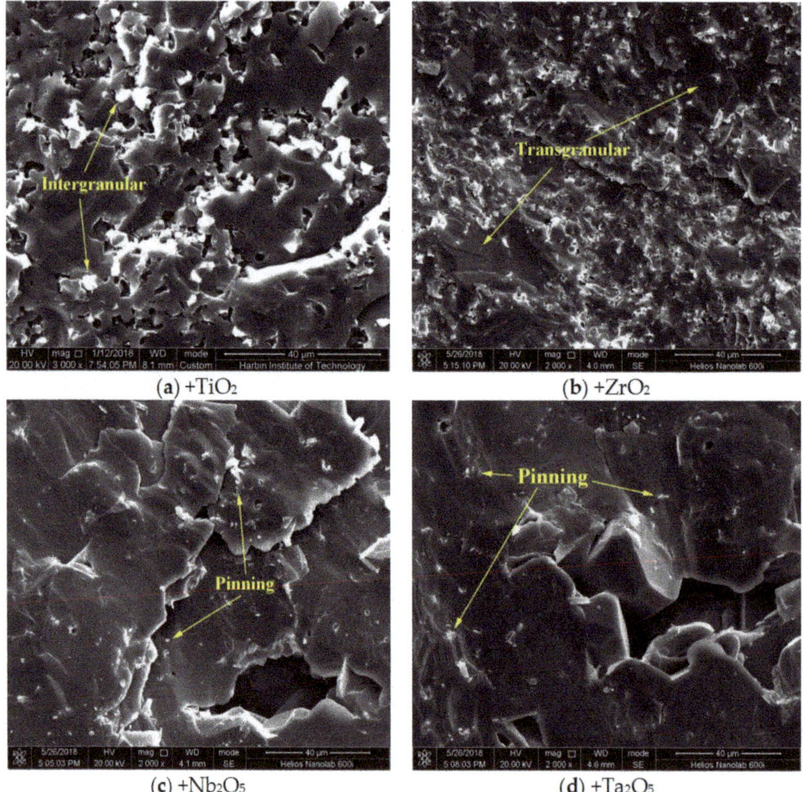

(**a**) +TiO_2

(**b**) +ZrO_2

(**c**) +Nb_2O_5

(**d**) +Ta_2O_5

Figure 4. The results of SEM analysis of sintered samples and the fracture surfaces of final products: (**a**) +TiO_2; (**b**) +ZrO_2; (**c**) +Nb_2O_5; (**d**) +Ta_2O_5.

The average flexural strength of the multiphase-ceramics samples prepared in this study was 321 MPa, while the lowest value was above 260 MPa. Compared with the reference strength values for pressureless sintering in the literature, the average strength value for the samples in this study was higher and was equivalent to those of hot press sintering samples.

One significant factor in the preparation of boron carbide ceramics is that they undergo toughening at high temperature. Therefore, it is necessary to find out how the final phase and microstructure influence the crack behavior. Figure 5 shows the crack propagation on the surfaces of the final samples after the hardness tests. In Figure 5a, crack bifurcation and deflection are apparent on the surface owing to significant aggregation among the second-phase particles of the TiO_2-added product, in which the crack propagation is effectively hindered by energy consumption. Figure 5b shows the cracks on the surface of the ZrO_2-added product, in which the crack deflection occurred without obvious presence of second-phase particles. This phenomenon can be analyzed using SEM images. In Figure 3b, the $(Zr, W)B_2$ second-phase particles are large and unevenly distributed. The cracks could pass directly through the B_4C matrix instead of being deflected if they did not meet the second phase. However, the fine $(Zr, W)B_2$ second-phase grains could cause strong crack deflection, making the addition of ZrO_2 an excellent to hinder crack propagation. In the SEM image of the Nb_2O_5-added sample, shown in Figure 5c, the second-phase particles had a moderate particle size compared to the above products, which were also uniformly distributed in the B_4C matrix. Additionally, it is obvious that the Nb_2O_5-added sample had higher fracture toughness owing to the crack deflection phenomenon. The morphology of the Ta_2O_5-added final product was similar to the Nb_2O_5-added final product according to Figure 5d. This was due to crack bifurcation, which hindered crack growth in both samples.

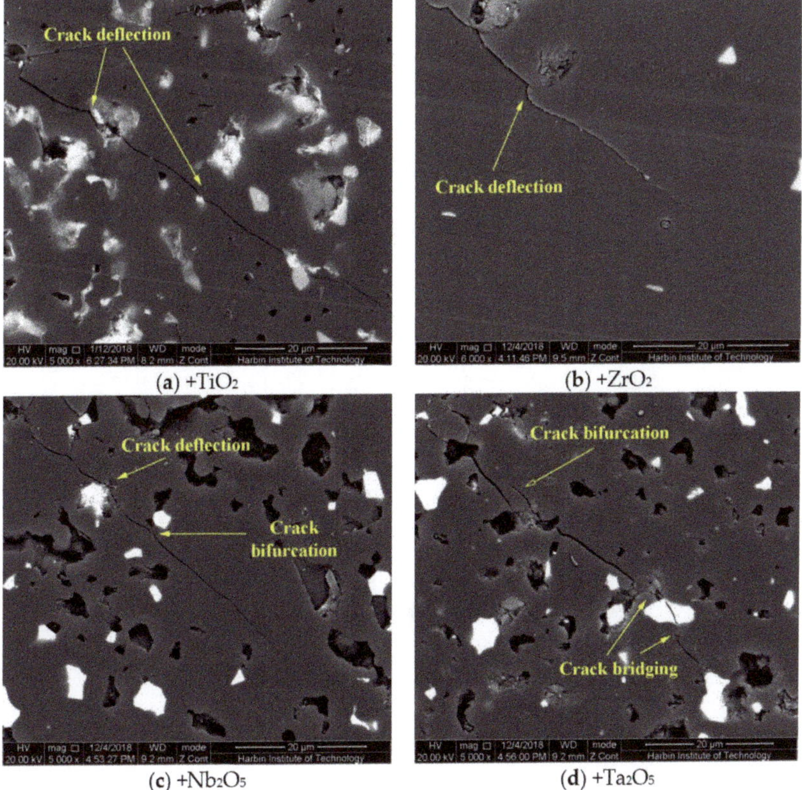

Figure 5. The results of SEM analysis of fracture surfaces and crack propagation of final products: (**a**) +TiO_2; (**b**) +ZrO_2; (**c**) +Nb_2O_5; (**d**) +Ta_2O_5.

The residual stresses were dispersed in the matrix around the second-phase particles. This phenomenon was always considered as the reason for the toughening of the composite ceramics. This kind of residual stress is, to a great extent, influenced by a mismatch in thermal expansion coefficients [41]. Different values mean that certain factors could induce microcracking behavior between the B_4C matrix and second-phase $(Me, W)B_2$ particles, which in turn could induce particle toughening.

The thermal expansion coefficient values for B_4C and the different MeB_x phases are shown in Table 5. Regarding Table 3, the CTE values of the second phases are higher than that of B_4C, except for in ZrB_2. Owing to the bigger disparities among the thermal expansion coefficients of the samples with added transition metal oxides and that of the B_4C matrix, we speculate that the residual stress for TiB_2, NbB_2, and TaB_2 could be higher than that of the other final products. This conclusion was proven considering the data in Table 4 and Figure 5. Except for the TiO_2-added sample, whose relative density was too low to reduce all of the mechanical properties, the Ta_2O_5-added and Nb_2O_5-added samples had a deflective crack path morphology and high fracture toughness. The thermal expansion coefficient mismatch between the B_4C matrix and the second phase MeB_2 or $(Me, W) B_x$ particles after the B_4C-MeB_2 polyphase ceramics were cooled down to room temperature resulted in the formation of residual stresses and microcracks. Second-phase particles with different thermal expansion coefficients can lead to differences in the residual stress among samples, which will affect the interface toughening during crack propagation.

Table 5. Comparison of thermal expansion coefficients (CTE) of boron carbide and the relevant transition metal borides [42,43].

Matter	B_4C	TiB_2	ZrB_2	NbB_2	TaB_2
CTE (ppm/K)	4.5	8.1	5.5	8.2	8.5

Additionally, the particle size, particle shape, particle orientation, surface energy, and interfacial bonding condition of the particles were determined to be the factors that could change the fracture toughness of the final products in this study. As shown in Figure 5, compared with the ZrO_2-added product (Figure 5b), the rest of the samples performed better in terms of the average second-phase particle distribution and having suitable particle sizes. Such results could help explain the higher fracture toughness values of these samples, as detailed in Table 4. As for the Nb_2O_5-added sample shown in Figure 5c, it can be seen that the crack was deflected rather than directly passing through upon encountering the second phase. This phenomenon could provide proof that the Nb_2O_5 additive had a significant second-phase toughening effect as it induced a high energy consumption. The fracture toughness of the Ta_2O_5-added sample was found to be slightly larger than that of the Nb_2O_5-added sample, whose cracks passed through the second phase. The energy from the crack propagation was largely consumed in this sample, with bifurcation and deflection occurring frequently. It was proven that Ta_2O_5 and Nb_2O_5 additives had a significant second-phase toughening effect.

The chemical reaction paths during the sintering process were similar in different B_4C-MeO_x samples. The pore distribution, particle size, and fracture modes were dependent on the metal oxide added (Me = Ti, Zr, Nb, or Ta). The different final phases and reaction processes led to differences among the B_4C-MeB_2 ceramics. The mechanical properties of the MeO_x-added samples also varied significantly, since the mechanical properties are closely related to the final microstructure. Among all of the samples, the microtopography of the Ta_2O_5-added sample stood out on the basis that it had the optimal mechanical properties. The B_4C-TaB_2 sample had good relative density and good distribution of the second phase, with a low number of pores. The fracture mode of this sample was a mixture of transgranular and intergranular fractures, in which the river-like patterns on the matrix and grain "pull-out" effect of the second-phase particles were observed. A significant toughening effect was caused by the evenly distributed second phase. Through the comprehensive analysis of the morphology and mechanical properties, it was seen that the second phase distribution of the B_4C-TaB_2 sample was uniform, and that the pores were fine and dispersed. The fine matrix and second-phase particles formed the interface, enhancing the toughness and making this the best-performing sample of the three. Meanwhile, the performance of the Nb_2O_5-added sample was similar to the Ta_2O_5-added sample regarding the mechanical properties and micromorphology. It is worth mentioning that there are few academic publications covering B_4C-NbB_2 and B_4C-TaB_2 multiphase ceramics. However, the comprehensive performance of the samples in this research was considerable. For example, the structure of the interface between the second phase and the matrix and its effect on the microstructure and mechanical properties could be further studied by TEM (Transmission electron microscope). At the same time, it could be possible to improve performance by growing or coating the Nb or Ta compounds on the surface.

4. Conclusions

Little research is available on how different transition metal oxides influence the behavior of B4C-based ceramics, especially for Ta_2O_5 and Nb_2O_5. It is of significance to find out the differences and rules among them. In this study, different transition metal oxides, such as TiO_2, ZrO_2, Nb_2O_5, and Ta_2O_5, were added to B_4C. The sintering process used was pressureless sintering at 2250 °C with 1 h holding time. By using FactSage software, the XRD pattern analysis shows that the reaction between B_4C and MeO_x is feasible, and then by measuring the relative density, SEM analysis, and measuring mechanical properties, the main conclusions are summarized. After adding MeO_x, the mechanical properties of boron carbide materials were observed to improve overall. Compared with other reports of B_4C ceramics, these ceramics have excellent properties. Among all of the investigated B_4C-MeB_2 multiphase ceramics, the main final phases are boron carbide and metal boride. Under the technological conditions of this study, all the samples have fine microstructure, few pores and uniformed second phase. B_4C-TaB_2 sample and B_4C-NbB_2 sample are due to the significant agglomeration of the second phase particles, and the crack deflection caused by the second phase grains led to the sample having better fracture toughness than others. The B_4C-TaB_2 sample had the most comprehensive properties. Its elastic modulus was 312.0 GPa, its hardness was 16.3 GPa, its flexural strength was 313.0 MPa, and its fracture toughness was 6.08 $MPa \cdot m^{1/2}$. The Nb_2O_5-added sample performed similarly—the comprehensive mechanical properties were better than the reported values when the mass fraction of second phase was around 5%. B_4C-NbB_2 and B_4C-TaB_2 multiphase ceramics could highlight a new direction of research related to the microstructures and mechanical properties among boron carbide composite ceramics. For example, further studies could explore the different influences of Ta, TaB_2, and TaC on B_4C-based ceramics, and could also investigate the optimal solution needed to improve performance and study the mechanisms behind performance differences.

Author Contributions: Conceptualization, G.L. and Y.Z.; methodology, G.L.; software, S.C.; validation, Y.Z., Y.F., and Y.W.; formal analysis, G.L., S.C., and Y.W.; investigation, S.C.; resources, Y.F. and Y.W.; data curation, S.C.; writing—original draft preparation, G.L.; writing—review and editing, S.C.; visualization, Y.F.; supervision, Y.W.; project administration, Y.W. All authors have read and agreed to the published version of the manuscript.

Funding: This research was funded by the National Natural Science Foundation of China under Projection No.5187206.

Conflicts of Interest: The authors declare no conflict of interest.

References

1. Vas, N.; Lazzari, R.; Besson, J.M.; Baroni, S.; Corso, A.D. Atomic structure and vibrational properties of icosahedral α-boron and B_4C boron carbide. *Comp. Mater. Sci.* **2000**, *17*, 127–132.

2. Zorzi, J.E.; Perottoni, C.A.; Jornada, J.A.H.D. Hardness and wear resistance of B_4C ceramics prepared with several additives. *Mater. Lett.* **2005**, *59*, 2932–2935. [CrossRef]

3. Hayun, S.; Weizmann, A.; Dariel, M.P. Microstructural evolution during the infiltration of boron carbide with molten silicon. *J. Eur. Ceram. Soc.* **2010**, *30*, 1007–1014. [CrossRef]

4. With, G.D. High temperature fracture of boron carbide: Experiments and simple theoretical models. *J. Mater. Sci.* **1984**, *19*, 457–466. [CrossRef]

5. Bejarano, G.G.; Caicedo, J.C.; Prieto, P.; Balogh, A.G. Cutting tool performance enhancement by using a $B_4C/BCN/C$-BN multilayer system. *Phys. Status Solidi C* **2007**, *4*, 4282–4287. [CrossRef]

6. Sun, J.L.; Liu, C.X.; Liu, J.X. Effect of Grain Size on Erosion Wear of $B_4C/TiC/Al_2O_3$ Ceramic Nozzles. *Appl. Mech. Mater.* **2015**, *740*, 32–35. [CrossRef]

7. Guo, T.; Wang, C.; Dong, L.M.; Liang, T.X. Preparation and Properties of B_4C/Graphite Neutron Absorption Ball. *Key Eng. Mater.* **2014**, *602*, 248–251. [CrossRef]

8. Kumar, S.; Sairam, K.; Sonber, J.K.; Murthy, T.S.R.C.; Reddy, V.; Nageswara, G.V.S.; Srinivasa, T. Hot-pressing of $MoSi_2$ reinforced B_4C composites. *Ceram. Int.* **2014**, *40*, 16099–16105. [CrossRef]

9. Radev, D.; Avramova, I.; Kovacheva, D.; Gautam, D.; Radev, I. Synthesis of Boron Carbide by Reactive-Pulsed Electric Current Sintering in the Presence of Tungsten Boride. *Int. J. Appl. Ceram. Tec.* **2016**, *13*, 997–1007. [CrossRef]

10. Huang, S.; Kim, V.; Jozef, V. In-situ synthesis and densification of B_4C-(Zr, Ti)B_2 composites by pulsed electric current sintering. *J. Chinese Ceram. Soc.* **2014**, *1*, 113–121.

11. He, R.; Jing, L.; Qu, Z.; Zhou, Z.; Ai, S.; Kai, W. Effects of ZrB_2 contents on the mechanical properties and thermal shock resistance of B_4C-ZrB_2 ceramics. *Mater. Design* **2015**, *71*, 56–61. [CrossRef]

12. Zou, J.; Huang, S.G.; Vanmeensel, K.; Zhang, G.J.; Vleugels, J.; Vander-Biest, O. Spark Plasma Sintering of Superhard B_4C-ZrB_2 Ceramics by Carbide Boronizing. *J. Am. Ceram. Soc.* **2013**, *96*, 1055–1059. [CrossRef]

13. Zhang, X.R.; Zhang, Z.X.; Wang, W.X.; Che, H.W.; Zhang, X.L.; Bai, Y.M.; Zhang, L.; Fu, Z.G. Densification behaviour and mechanical properties of B_4C-SiC intergranular/intragranular nanocomposites fabricated through spark plasma sintering assisted by mechanochemistry. *Ceram. Int.* **2017**, *43*, 1904–1910. [CrossRef]

14. You, Y.; Tan, D.W.; Guo, W.M.; Wu, S.H.; Lin, H.T.; Luo, Z. TaB_2 powders synthesis by reduction of Ta_2O_5 with B_4C. *Ceram. Int.* **2017**, *43*, 897–900. [CrossRef]

15. Saeedi, H.M.; Baharvandi, H.R. Comparing the effects of different sintering methods for ceramics on the physical and mechanical properties of B_4C-TiB_2 nanocomposites. *Int. J. Refract. Met. H* **2015**, *51*, 224–232. [CrossRef]

16. Zhang, X.R.; Zhang, Z.X.; Wang, W.M.; Zhang, X.L.; Fu, Z.Y. Preparation of B_4C composites toughened by TiB_2-SiC agglomerates. *J. Eur. Ceram. Soc.* **2017**, *37*, 865–869. [CrossRef]

17. Husarova, I.O.; Potapov, O.M.; Solodkyi, I.V.; Bogomol, I.I. Production and Properties of B_4C-TiB_2 Composites with Isotropic Eutectic Microstructure. *Powder Metall. Met. Ceram.* **2018**, *57*, 209–214. [CrossRef]

18. Wang, G.F.; Zhang, J.H.; Zhang, C.; Zhang, K.F. Densification and Mechanical Properties of B_4C Based Composites Sintered by Reaction Hot-Pressing. *Key Eng. Mater.* **2010**, *434*, 24–27. [CrossRef]

19. Tamari, N.; Kobayashi, H.; Tanaka, T.; Kondoh, I.; Kose, S. Mechanical Properties of B_4C-SiC Whisker Composite Ceramics. *J. Ceram. Soc. Jpn.* **2010**, *98*, 1159–1163. [CrossRef]

20. Jiang, T.; Tian, C.C. Investigation of Microstructure and Thermal Shock Resistance of the B_4C/BN Composites Fabricated by Hot-Pressing Process. *Key Eng. Mater.* **2012**, *512*, 748–752. [CrossRef]

21. Yan, Z.; Liu, J.; Zhang, J.; Ma, T.; Li, Z.; Yan, Z.; Jie, L. Biomorphic silicon/silicon carbide ceramics from birch powder. *Ceram. Int.* **2011**, *37*, 725–730. [CrossRef]

22. Neuman, E.W.; Hilmas, G.E.; Fahrenholtz, W.G. Processing, microstructure, and mechanical properties of zirconium diboride-boron carbide ceramics. *Ceram. Int.* **2017**, *43*, 6942–6948. [CrossRef]

23. Neuman, E.W.; Brown-Shaklee, H.J.; Hilmas, G.E.; Fahrenholtz, W.G. Titanium diboride-silicon carbide-boron carbide ceramics with super-high hardness and strength. *J. Am. Ceram. Soc.* **2017**, *101*, 497–501. [CrossRef]

24. Liu, L.; Geng, G.; Jiang, Y.; Wang, Y.; Hai, W.; Sun, W.Z.; Chen, Y.H.; Wu, L. Microstructure and mechanical properties of tantalum carbide ceramics: Effects of Si_3N_4 as sintering aid. *Ceram. Int.* **2017**, *43*, 5136–5144. [CrossRef]

25. Zhang, C.; Gupta, A.; Seal, S.; Boesl, B.; Agarwal, A. Solid solution synthesis of tantalum carbide-hafnium carbide by spark plasma sintering. *J. Am. Ceram. Soc.* **2017**, *100*, 1853–1862. [CrossRef]

26. Zhang, W.; Yamashita, S.J.; Kita, H. Progress in pressureless sintering of boron carbide ceramics—A review. *Adv. Appl. Ceram.* **2019**, *118*, 222–239. [CrossRef]

27. Gao, D.; Jing, J.; Yu, J.; Guo, X.; Zhang, Y.; Gong, H.; Zhang, Y. Graphene platelets enhanced pressureless sintered B_4C ceramics. *Roy. Soc. Open Sci.* **2018**, *5*, 171837. [CrossRef]

28. Moshtaghioun, B.M.; Diego, G.G.; Arturo, D.R. High-temperature plastic deformation of spark plasma sintered boron carbide-based composites: The case study of B_4C-SiC with/without graphite (g). *J. Eur. Ceram. Soc.* **2016**, *36*, 1127–1134. [CrossRef]

29. Ariza Galván, E.; Montealegre-Meléndez, I.; Arévalo, C.; Kitzmantel, M.; Neubauer, E. Ti/B$_4$C Composites Prepared by In Situ Reaction Using Inductive Hot Pressing. *Key Eng. Mater.* **2017**, *742*, 121–128. [CrossRef]

30. Sairam, K.; Vishwanadh, B.; Sonber, J.K.; Murthy, T.S.R.C.; Majumdar, S.; Mahata, T.; Basu, B. Competition between densification and microstructure development during spark plasma sintering of B$_4$C-Eu$_2$O$_3$. *J. Am. Ceram. Soc.* **2017**, *101*, 2516–2526. [CrossRef]

31. Perevislov, S.N.; Lysenkov, A.S.; Vikhman, S.V. Effect of Si additions on the microstructure and mechanical properties of hot-pressed B$_4$C. *Inorg. Mater.* **2017**, *53*, 376–380. [CrossRef]

32. Yue, X.Y.; Zhao, S.M.; Yu, L.; Ru, H.Q. Microstructures and Mechanical Properties of B$_4$C-TiB$_2$ Composite Prepared by Hot Pressure Sintering. *Key Eng. Mater.* **2010**, *434*, 50–53. [CrossRef]

33. Uygun, B.; Göller, G.; Onüralp, Y.; Şahin, F.Ç. Production and Characterization of Boron Carbide-Titanium Diboride Ceramics by Spark Plasma Sintering Method. *Adv. Sci. Technol.* **2010**, *63*, 68–73. [CrossRef]

34. Baharvandi, H.R.; Hadian, A.M.; Alizadeh, A. Processing and Mechanical Properties of Boron Carbide-Titanium Diboride Ceramic Matrix Composites. *Appl. Compos. Mater.* **2006**, *13*, 191–198. [CrossRef]

35. Xu, C.; Cai, Y.; Lodström, K.; Li, F.Z.; Esmaeilzadeh, S.; Zhang, G.-J. Spark plasma sintering of B$_4$C ceramics: The effects of milling medium and TiB$_2$ addition. *Int. J. Refract. Met. Hard Mater.* **2012**, *30*, 39–144. [CrossRef]

36. Dudina, D.V.; Hulbert, D.M.; Jiang, D.; Unuvar, C.; Cytron, S.J.; Mukherjee, A.K. In situ boron carbide-titanium diboride composites prepared by mechanical milling and subsequent spark plasma sintering. *J. Mater. Sci.* **2008**, *43*, 3569–3576. [CrossRef]

37. Liu, A.D.; Qiao, Y.J.; Liu, Y.Y. Pressureless sintering and properties of boron carbide-titanium diboride composites by in situ reaction. *Key Eng. Mater.* **2013**, *525*, 321–324. [CrossRef]

38. Demirskyi, D.; Sakka, Y.; Vasylkiv, O. High-Strength B$_4$C-TaB$_2$ Eutectic Composites Obtained via In Situ by Spark Plasma Sintering. *J. Am. Ceram. Soc.* **2016**, *99*, 2436–2441. [CrossRef]

39. Liu, R.; Ru, H.Q.; Zhao, Y.; Tang, D. In situ synthesis of B$_4$C ceramics toughened by ZrB$_2$ particles. *Chin. J. Mater. Res.* **2006**, *20*, 611–616.

40. Ho, C.J.; Tuan, W.H. Toughening and strengthening zirconia through the addition of a transient solid solution additive. *J. Eur. Ceram. Soc.* **2012**, *32*, 335–341. [CrossRef]

41. Liu, G.Q.; Chen, S.X.; Zhao, Y.W.; Fu, Y.D.; Wang, Y.J. The effect of transition metal carbides MeC (Me = Ti, Zr, Nb, and W) on mechanical properties of B$_4$C ceramics fabricated via pressureless sintering. *Ceram. Int.* **2020**, *46*, 27283–27291. [CrossRef]

42. Awaji, H.; Choi, S.M.; Yagi, E. Mechanisms of toughening and strengthening in ceramic-based nanocomposites. *Mech. Mater.* **2002**, *34*, 411–422. [CrossRef]

43. Vahldiek, F.W. Electrical resistivity, elastic modulus, and debye temperature of titanium diboride. *J. Less Common Met.* **1967**, *12*, 202–209. [CrossRef]

Publisher's Note: MDPI stays neutral with regard to jurisdictional claims in published maps and institutional affiliations.

© 2020 by the authors. Licensee MDPI, Basel, Switzerland. This article is an open access article distributed under the terms and conditions of the Creative Commons Attribution (CC BY) license (http://creativecommons.org/licenses/by/4.0/).

 *coatings*

Article

Insights in to the Electrochemical Activity of Fe-Based Perovskite Cathodes toward Oxygen Reduction Reaction for Solid Oxide Fuel Cells

Dan Ma, Juntao Gao, Tian Xia, Qiang Li *, Liping Sun, Lihua Huo and Hui Zhao

Key Laboratory of Functional Inorganic Material Chemistry, Ministry of Education, School of Chemistry and Materials Science, Heilongjiang University, Harbin 150080, China; madan18845166284@hotmail.com (D.M.); 20b911032@stu.hit.edu.cn (J.G.); xiatian@hlju.edu.cn (T.X.); sunliping@hlju.edu.cn (L.S.); huolihua@hlju.edu.cn (L.H.); zhaohui@hlju.edu.cn (H.Z.)
* Correspondence: liqiang@hlju.edu.cn

Received: 27 November 2020; Accepted: 18 December 2020; Published: 19 December 2020

Abstract: The development of novel oxygen reduction electrodes with superior electrocatalytic activity and CO_2 durability is a major challenge for solid oxide fuel cells (SOFCs). Here, novel cobalt-free perovskite oxides, $BaFe_{1-x}Y_xO_{3-\delta}$ (x = 0.05, 0.10, and 0.15) denoted as BFY05, BFY10, and BFY15, are intensively evaluated as oxygen reduction electrode candidate for solid oxide fuel cells. These materials have been synthesized and the electrocatalytic activity for oxygen reduction reaction (ORR) has been investigated systematically. The BFY10 cathode exhibits the best electrocatalytic performance with a lowest polarization resistance of 0.057 Ω cm^2 at 700 °C. Meanwhile, the single cells with the BFY05, BFY10 and BFY15 cathodes deliver the peak power densities of 0.73, 1.1, and 0.89 W cm^{-2} at 700 °C, respectively. Furthermore, electrochemical impedance spectra (EIS) are analyzed by means of distribution of relaxation time (DRT). The results indicate that the oxygen adsorption-dissociation process is determined to be the rate-limiting step at the electrode interface. In addition, the single cell with the BFY10 cathode exhibits a good long-term stability at 700 °C under an output voltage of 0.5 V for 120 h.

Keywords: solid oxide fuel cells; oxygen reduction electrode; electrocatalytic activity

1. Introduction

Perovskite-type materials are the promising cathodes for solid oxide fuel cells (SOFCs) [1–3]. Among the perovskite oxides, the Co-containing perovskite materials with mixed ionic-electronic conducting feature have been widely investigated due to their remarkable electrochemical performance for ORR [4–6]. However, Co-containing oxides show some other drawbacks, such as poor chemical stability, higher thermal expansion coefficient, and strong volatility, which inhibit their wide applications in SOFCs [7,8]. To address these issues, developing new cathodes with improved electrochemical performance and good chemical stability is an important trend. Recently, Fe-based perovskite materials exhibit attractive chemical compatibility and excellent electrocatalytic activity, such as $SrFe_{1-x}Ti_xO_{3-\delta}$, $SrFe_{0.8}Nb_{0.2-x}Ta_xO_{3-\delta}$, $La_{1-x}Sr_xFeO_{3-\delta}$, and $Ba_{1-x}La_xFeO_{3-\delta}$ [9–12].

Among the Fe-based oxides, $BaFeO_{3-\delta}$ presents attractive oxygen permeation flux and fast oxygen surface exchange kinetics [13,14]. This is mainly due to variable valence and excellent chemical stability of Fe ions, as well as lower valence and larger ionic radius of Ba^{2+}, which facilitates the electrochemical performance and oxygen transport of the materials [15]. Cation-doping is commonly adopted to stabilize the cubic lattice of $BaFeO_{3-\delta}$ with disordered oxygen vacancies, such as La^{3+}, Sm^{3+}, and Ca^{2+} in the A site and Nb^{5+}, Sn^{4+}, In^{3+}, and Ni^{2+} in the B site [12,16–18]. Lu et al. found that In^{3+} doping in the B site of $BaFeO_{3-\delta}$ can enhance oxygen permeation flux, in which $BaFe_{0.9}In_{0.1}O_{3-\delta}$ presented the higher oxygen permeation flux of 1.11 mL cm^{-2} min^{-1} at 950 °C [19]. Song et al. reported an attractive cathode candidate of $BaFe_{1-x}Bi_xO_{3-\delta}$ for SOFCs. The $BaFe_{0.9}Bi_{0.1}O_{3-\delta}$ cathode exhibited a lower polarization resistance of 0.133 Ω cm^2 at 750 °C and a high oxygen vacancy concentration of 0.408. However, the thermal expansion coefficient of $BaFe_{0.9}Bi_{0.1}O_{3-\delta}$ is very large (26.697 $\times$ 10^{-6} K^{-1}). Additionally, the oxygen permeability and oxygen non-stoichiometry of $BaFe_{1-x}Y_xO_{3-\delta}$ have been investigated by Liu et al. They found that Y-doping promotes the oxygen vacancy concentration and oxygen ion migration, which is believed to be favorable for transport of the oxygen ion in the bulk electrode, thereby resulting in the enhanced electrocatalytic performance [20].

In this work, Fe-based perovskite $BaFe_{1-x}Y_xO_{3-\delta}$ oxides have been investigated as the prominent cathodes for SOFCs. The crystalline structure, CO_2 tolerance, and electrocatalytic activity for ORR of the $BaFe_{1-x}Y_xO_{3-\delta}$ cathodes are systematically investigated. The results provide an effective strategy for designing novel cathode electrocatalysts for SOFCs.

2. Experimental

2.1. Material Preparation

The $BaFe_{1-x}Y_xO_{3-\delta}$ (x = 0.05–0.15) samples were synthesized by the solid-state reaction. Stoichiometric amounts of $BaCO_3$ (99.99%, Tianjin Guangfu Co. Ltd., Tianjin, China), Fe_2O_3 (99.99%, Tianjin Guangfu Co. Ltd.), and Y_2O_3 (99.99%, Tianjin Guangfu Co. Ltd.) were mixed and ground via the ball milling using ethanol as the dispersant. Afterwards, the mixture was pre-fired at 1000 °C for 10 h with a heating rate of 5 °C min^{-1} in air and then re-milled for 1 h, followed by calcining at 1300 °C for 12 h to obtain the final products.

2.2. Characterization

The crystal structure of BFYx cathodes were identified using X-ray diffraction (Bruker D8 advance) with filtered Cu-Kα radiation (λ = 1.5148 Å) source in a 2θ range of 10°–80°. The XRD data was analyzed to obtain the structural parameters by using Rietveld refinement method using the Rietica software (Version 1.7.7.8) program. The oxygen desorption property of the cathode catalysts was carried out by O_2 temperature-programmed desorption (O_2-TPD) with the TP-5076 instrument (Tianjin Xianquan, Co. Ltd., Tianjin, China). The electrical conductivity was measured between 100 and 800 °C in air by standard four-probe DC method with a Keithley 2400 SourceMeter Keithley Instruments Inc., Cleveland, OH, USA). The oxygen non-stoichiometry of the samples at room temperature was determined with the iodometric titration method, as described elsewhere [21,22]. The oxygen non-stoichiometry at high temperature was measured by thermogravimetric analysis (TGA, SETARAM, Caluire et Cuire, France).

2.3. Electrochemical Test

The dense $Ce_{0.9}Gd_{0.1}O_{1.95}$ (CGO) electrolyte was fabricated by pressing CGO powders (SOFCMAN Co. Ltd., Ningbo, China) uniaxially at 220 MPa, and sintered at 1450 °C for 24 h. For the fabrication of symmetrical cells (BFYx|CGO|BFYx), the BFYx electrode powders were mixed with terpineol and ethyl cellulose to prepare the cathode slurry. The slurry was symmetrically coated on the $Ce_{0.9}Gd_{0.1}O_{2-\delta}$ (CGO) electrolyte and sintered at 900 °C for 4 h. The electrochemical impedance spectroscopy (EIS) of symmetric cell was acquired by an Autolab PGSTAT302N workstation in the frequency range of 10^{-2}–10^6 Hz under open voltage conditions at 500–700 °C. To explore the electrochemical process for ORR of the electrode, the EIS spectra were collected under different oxygen partial pressure (PO_2).

The anode-supported half cell (NiO-YSZ|YSZ|CGO) was bought from Ningbo SuoFuRen Energy Co. Ltd. (Ningbo, China). The BFYx cathode slurry was printed onto the CGO barrier layer, and subsequently co-fired at 900 °C for 4 h. The electrochemical performance of anode-supported fuel cells was tested using electrochemical workstation (ZAHNER, IM6e, Kronach, Germany. The single cell was mounted on the alumina tube, while the humidified H_2 (3% H_2O) with a flow rate of 80 mL min^{-1} and ambient air were used as the fuel and oxidant, respectively.

3. Results and Discussion

Figure 1a displays the XRD patterns of BFYx samples. The diffraction profiles reveal that the BFYx oxides crystallize in a cubic perovskite structure with *Pm-3m* space group. The magnified XRD patterns between 29 and 33° are presented in Figure 1a. The diffraction peaks gradually shift to a lower angle direction with increasing the doping fraction, indicating the expansion of the lattice constants. To further obtain the lattice constants of the materials, Rietveld refined XRD data of BFY10 samples are given in Figure 1b. The refined lattice constants of BFYx samples are summarized in Table 1. The increase in cell volume from 67.220 Å^3 ($x = 0.05$) to 69.148 Å^3 ($x = 0.15$) is identified, which is attributed to the larger ionic radius of Y^{3+} (0.90 Å) relative to that of Fe^{3+} (0.55 Å). This phenomenon indicates that the Fe sites in BaFeO$_3$ are partially replaced by the Y ions. Furthermore, to examine the chemical compatibility between electrode and electrolyte, the mixtures of BFYx and CGO were co-fired at 1000 °C for 12 h. Figure 1c presents the XRD patterns of the calcined BFYx-CGO mixtures. No obvious impurities can be detected, revealing the favorable chemical compatibility of the BFYx cathodes with CGO electrolyte at 1000 °C.

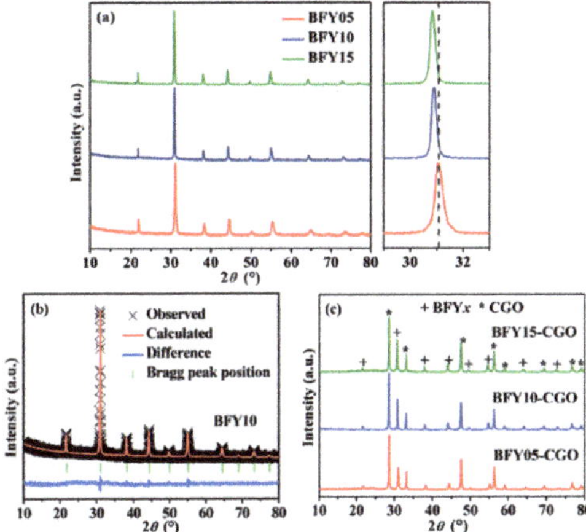

Figure 1. (**a**) Room temperature XRD patterns of BFYx samples and the magnified view of the XRD patterns, 2θ = 29 to 33°; (**b**) Rietveld refinement plot of BFY10 XRD data; (**c**) XRD patterns of BFYx-CGO composites fired at 1000 °C for 12 h.

Table 1. Lattice parameters of BFYx samples.

Space Group	$x = 0.05$	$x = 0.1$	$x = 0.15$
	Pm-3m	*Pm-3m*	*Pm-3m*
a (Å)	4.066	4.091	4.104
b (Å)	4.066	4.091	4.104
c (Å)	4.066	4.091	4.104
V (Å^3)	67.220	68.468	69.148
χ^2	2.171	1.906	4.190
R_{wp} (%)	7.762	7.664	9.811
R_P (%)	5.975	6.023	7.878

The oxygen mobility and reducibility of Fe ions for the BFYx samples were studied by O_2-TPD. Figure 2a presents the representative O_2-TPD profiles of the BFYx samples. The curves show two desorption peaks at around 200 and 650 °C in all samples. The first broad peak located at ~200–300 °C is associated with the desorption process of the chemisorbed oxygen on the material surface. The second oxygen desorption peak may be ascribed to the reduction of Fe^{4+} to Fe^{3+} at 300–650 °C [23]. It is noteworthy that the BFY10 material shows the largest area of second desorption peak among the BFYx samples, indicating the highest oxygen vacancy concentration and favorable oxygen mobility of the BFY10. The oxygen non-stoichiometry (δ) of the BFYx samples at elevated temperatures was explored by TGA in air, as presented in Figure 2b. The δ values were determined by TGA results and the initial oxygen non-stoichiometry (δ_0) values at room temperature were obtained by the iodometric titration. The initial weight loss from room temperature to 300 °C is associated with the evaporation of adsorbed water. When the temperature is above 300 °C, the obvious weight loss is due to the reduction of Fe^{4+} to Fe^{3+}. It can be seen that the δ values of the samples decreases with the doping content from room temperature to 600 °C. However, when increasing the temperature to 600 °C, the BFY10 possesses the largest oxygen vacancy concentration, meaning its excellent oxygen ions mobility and promoted catalytic activity for ORR.

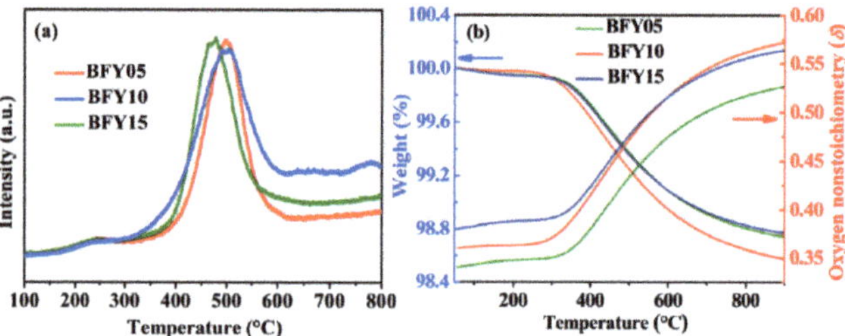

Figure 2. (a) TPD curves of BFYx samples measured from 100 to 800 °C; (b) TG curves and oxygen non-stoichiometry (δ) of BFYx between 50 and 900 °C.

The temperature dependence of electrical conductivity for the BFYx samples within the temperature range of 100–800 °C in air is presented in Figure 3a. The electrical conductivity of all samples shows a similar trend as a function of temperature. When increasing the temperature, the electrical conductivity of BFYx initially increases and reaches a maximum value at about 400 °C, and subsequently decreases between 400 and 800 °C. This indicates a transition from semi-conducting behavior to metal-like conduction. In addition, it is observed that the electrical conductivity gradually decreases with increasing Y-doping fraction. This phenomenon may be due to the higher Y content leads to the reduction of Fe^{4+} to Fe^{3+} or Fe^{2+} for the charge compensation, resulting in a decrease in the electrical conductivity. The maximum values of electrical conductivity are 9.81, 5.03, and 1.67 S cm^{-1} for BFY05, BFY10 and BFY15, respectively. These values are comparable to those of other reported $BaFeO_{3-\delta}$-based cathodes, such as $BaFe_{0.95}Nb_{0.05}O_{3-\delta}$ (9.5 S cm^{-1}) [16], $BaFe_{0.95}Zr_{0.05}O_{3-\delta}$ (7.5 S cm^{-1}) [14] and $BaFe_{0.8}In_{0.2}O_{3-\delta}$ (2.3 S cm^{-1}) [19]. Furthermore, the Arrhenius plots of electrical conductivity are presented in Figure 3b. The calculated activation energies (E_a) of BFYx are 0.29, 0.38, and 0.43 eV for BFY05, BFY10 and BFY15, respectively.

The EIS spectra were used to demonstrate electrocatalytic performance of the symmetric cells of BFYx|CGO|BFYx. Figure 4a displays the Nyquist plots of the BFYx cathodes at 700 °C in air. In general, the polarization resistance (R_p) value of the electrode is a crucial descriptor for the cathode performance, and the lower R_p value means a superior electrochemical activity for ORR. The R_p of BFY05, BFY10 and BFY15 cathodes are 0.136, 0.057, and 0.107 Ω cm^2 at 700 °C, respectively, suggesting highly electrocatalytic performance of the BFYx cathodes. The R_p value (700 °C) of BFY10 cathode is smaller than that of the Fe-based perovskite electrodes (Figure 4b) [24–27]. Furthermore, the Arrhenius plots of the polarization resistance are presented in Figure 4c. The calculated E_a values are 1.40, 1.33 and 1.45 eV for BFY05, BFY10, and BFY15, respectively. Moreover, the BFY10 cathode presents the lowest E_a value, implying the highest electrocatalytic performance.

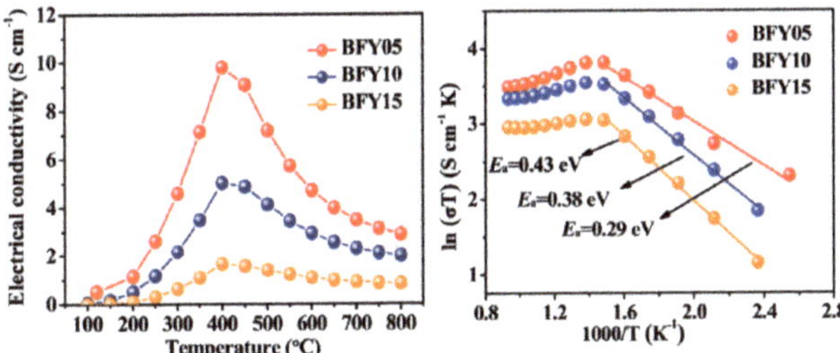

Figure 3. (**a**) The electrical conductivity of BFY*x* samples at 100–800 °C; (**b**) Arrhenius plots of BFY*x* with temperature.

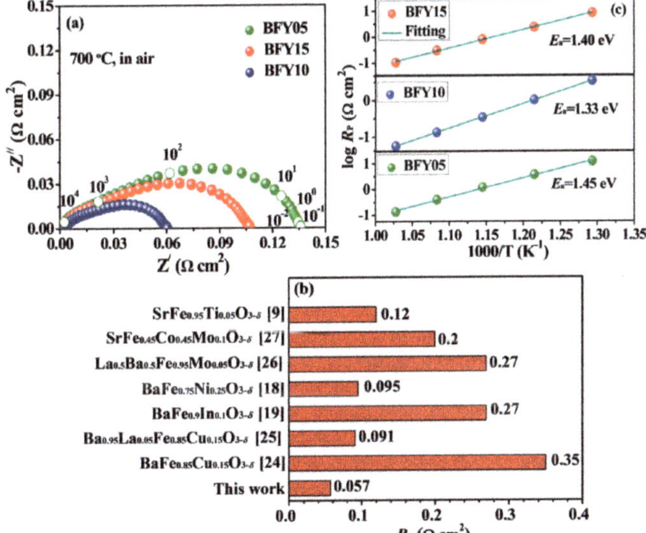

Figure 4. (**a**) Impedance spectra of BFY*x* measured at 700 °C; (**b**) Comparison of R_P for different Fe-based perovskite cathodes at 700 °C; (**c**) Arrhenius plots of R_p for BFY*x* cathodes.

To further clarify the electrochemical processes of the cathode, EIS spectra of the BFY10 electrode are systematically studied under varied Po_2 at 700 °C, as shown in Figure 5a. Clearly, the impedance spectra are consisted of three separable high-frequency, intermediate-frequency and low-frequency arcs, respectively, suggesting that the three different electrode processes occur on the cathode. The EIS data are further fitted with an equivalent circuit using the model of $[R_{ohm}\text{-}(R_H\text{-}CPE_H)\text{-}(R_M\text{-}CPE_M)\text{-}(R_L\text{-}CPE_L)]$ (inset in Figure 5a) and analyzed by the distribution of relaxation time (DRT) method. Figure 5b displays the DRT results of BFY10 cathode under different Po_2 at 700 °C. It can be seen that the typical DRT plots presents three distinct peaks, the high-frequency peak P1 (HF), and intermediate-frequency peak P2 (MF) and low-frequency peak P3 (LF), corresponding to charge transport process, adsorption-dissociation process of oxygen molecule, and gaseous diffusion, respectively [28,29]. Additionally, the relationship between R_p and PO_2 can be expressed by the following formula: $R_p = k(PO_2)^{-m}$ (1) [30,31]. The dependence of the R_{HF}, R_{MF} and R_{LF} for the BFY10 cathode on the PO_2 at 700 °C is presented in Figure 5c. One should note that the m values are 0.26, 0.48, and 1.03 in high-frequency, intermediate-frequency and low-frequency region, respectively, which are ascribed to the charge transfer process ($O_{ads.} + 2e^{/} + V_O^{\cdot\cdot} \Leftrightarrow O_O^x$, $m = 1/4$), the adsorption-dissociation of the oxygen molecule process ($O_{2,ads.} \Leftrightarrow 2O_{ads.}$, $m = 1/2$) and the adsorption and diffusion of gaseous oxygen on the electrode surface $(O_2(g) \Leftrightarrow O_{2ads.}, m = 1)$ [32]. Furthermore, it can be found that the R_{MF} is higher than R_{HF} and R_{LF}, implying that the rate-determining step for ORR is dominated by intermediate-frequency arc assigned to the oxygen adsorption-dissociation.

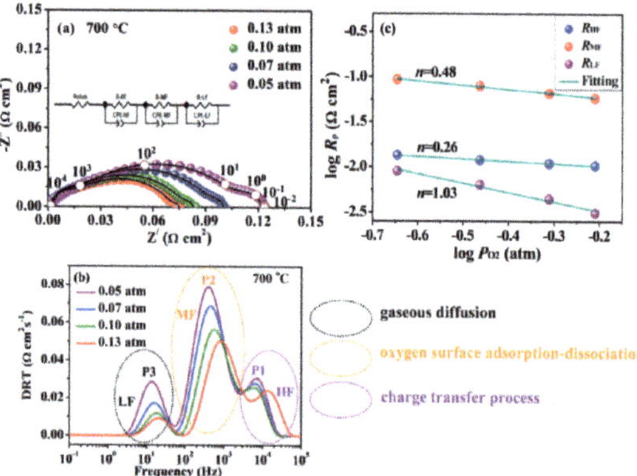

Figure 5. (**a**) Impedance spectra and (**b**) DRT results of BFY10 cathode under different oxygen partial pressure at 700 °C; (**c**) R_H, R_M, and R_L of BFY10 cathode as a function at 700 °C.

Some Ba-based cathodes for SOFCs show chemical instability under CO_2-containing atmospheres because of the formation of $BaCO_3$ on the cathode surface, diminishing oxygen reduction kinetics [33,34]. To evaluate the CO_2 tolerance of the electrode, EIS spectra of the BFY10 cathode were acquired in CO_2-containing air (3%, 5%, 10%) at 700 °C, as presented Figure 6a,b. It can be seen that the R_p value significantly increases with increasing the concentration of CO_2. However, the R_p recovers to the initial value after removing CO_2 atmosphere, which demonstrates the outstanding CO_2 tolerance of the BFY10 cathode. Generally, the average metal oxide binding energy (ABE) is normally used to assess the CO_2 durability of the cathode materials [35]. More negative ABE value indicates that oxide has excellent CO_2 tolerance. The ABE is calculated based on the following equation [36]:

$$ABE = ABE(A\text{-}O) + ABE(B\text{-}O) \tag{2}$$

$$ABE(A - O) = \frac{x_A}{12m}\left(\Delta_f H^o_{A_m O_n} - m\Delta H^o_A - \frac{n}{2}D_{O_2}\right) \tag{3}$$

$$ABE(B - O) = \frac{x_B}{6m}\left(\Delta_f H^o_{B_m O_n} - m\Delta H^o_B - \frac{n}{2}D_{O_2}\right) \tag{4}$$

where x_A and x_B are the molar fraction of A and B metals; ΔH^o_A and ΔH^o_B are the sublimation heat of A and B metals; $\Delta_f H^o_{A_m O_n}$ and $\Delta_f H^o_{B_m O_n}$ are the formation heat of $A_m O_n$ and $B_m O_n$ oxides, and D_{O2} is the dissociation energy of O_2. The ABE values of BFYx oxides are -287.18 kJ mol^{-1}, -291.35 kJ mol^{-1}, and -295.51 kJ mol^{-1}, respectively, which are higher than cobalt-free perovskite cathodes, such as $Ba_{0.5}Sr_{0.5}Co_{0.8}Fe_{0.2}O_{3-\delta}$ (-274 kJ mol^{-1}) [37], $Bi_{0.5}Sr_{0.5}FeO_{3-\delta}$ (-276.67 kJ mol^{-1}) and $Bi_{0.5}Sr_{0.5}Fe_{0.9}Ta_{0.1}O_{3-\delta}$ (-296.97 kJ mol^{-1}) [26]. These results confirm that the BFYx cathodes have satisfactory CO_2 tolerance.

To further demonstrate the practical application of the BFYx cathodes, the single cells were fabricated and tested. Figure 7a–c displays the *I–V* and *I–P* curves of the single cells with the BFYx cathodes 600–700 °C using humidified hydrogen and air as the fuel and oxidant, respectively. At 700 °C, the peak powder densities of 0.73, 1.1, and 0.89 W cm^{-2} are achieved in the BFY05, BFY10 and BFY15 cathodes, respectively. It should be noted that the single cell with the BFY10 cathode shows the highest peak powder density among the BFYx electrodes, which is associated with superior electrochemical performance for ORR. In addition, the peak powder density of single cell with the BFY10 cathode is higher than those of other cobalt-free perovskite cathodes [38–44]. Furthermore, Figure 7d displays the operating stability of the fuel cell with the BFY10 cathode during a period of 120 h. The single cell presents a stable current density and peak power density under an output voltage of 0.5 V without noticeable attenuation, suggesting that the BFY10 electrode has outstanding durability during the operating process. The remarkable electrocatalytic properties indicate that the BFY10 oxide is a highly attractive cathode candidate for SOFCs.

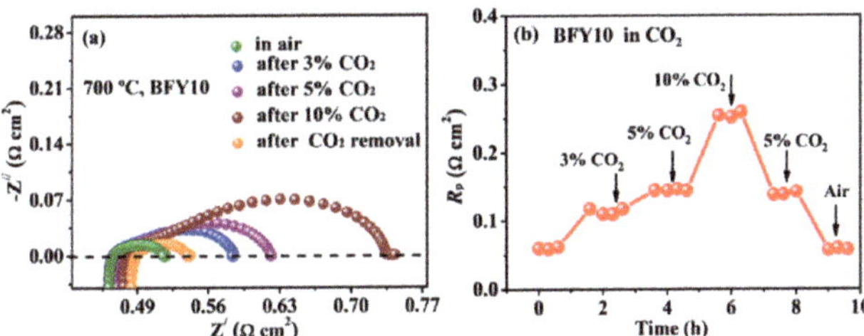

Figure 6. (a) Impedance spectra and (b) short-term stability of BFY10 cathode in various CO_2 concentrations at 700 °C.

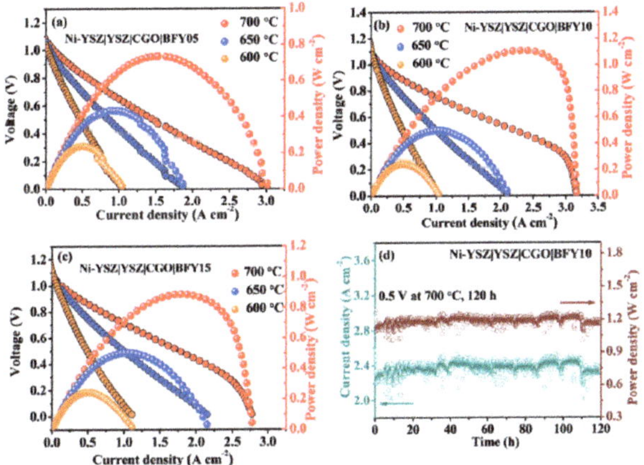

Figure 7. (a–c) I–V and I–P curves of the single cells with BFYx cathodes at 600–700 °C; (d) Stability test of the single cell with BFY10 cathode under an output voltage of 0.5 V.

4. Conclusions

In summary, Fe-based perovskite $BaFe_{1-x}Y_xO_{3-\delta}$ oxides with excellent ORR performance and CO_2 durability are evaluated as the cathode materials for SOFCs. Benefiting from cubic perovskite structure and large oxygen vacancy concentration, the BFY10 cathode presents outstanding electrochemical activity for ORR with a lower R_P value of 0.057 Ω cm^2 at 700 °C. In addition, the single fuel with the BFY10 cathode delivers a peak powder density of 1.1 W cm^{-2} at 700 °C, along with negligible attenuation over a period of 120 h. Furthermore, DRT study verifies that the adsorption-dissociation of the oxygen molecule process is the rate-limiting step on the cathode.

Author Contributions: D.M., Q.L., and H.Z. conceived and designed the experiments, and performed the experiments; J.G. and T.X. analyzed the data; L.S. and L.H. contributed reagents/materials/analysis tools; D.M. and Q.L. wrote the paper. All authors have read and agreed to the published version of the manuscript.

Funding: The project was supported by National Natural Science Foundation of China (51672072, 51972100) and Heilongjiang Provincial Fund for Distinguished Young Scholars (JC2018014).

Conflicts of Interest: The authors declare no conflict of interest.

References

1. Fan, L.; Zhu, B.; Su, P.-C.; He, C. Nanomaterials and technologies for low temperature solid oxide fuel cells: Recent advances, challenges and opportunities. *Nano Energy* **2018**, *45*, 148–176. [CrossRef]
2. Choudhury, A.; Chandra, H.; Arora, A. Application of solid oxide fuel cell technology for power generation—A review. *Renew. Sustain. Energy Rev.* **2013**, *20*, 430–442. [CrossRef]
3. Steele, B.C.H.; Heinzel, A. Materials for fuel-cell technologies. *Nat. Cell Biol.* **2001**, *414*, 345–352. [CrossRef]
4. Liu, P.; Luo, Z.F.; Kong, J.R.; Yang, X.F.; Liu, Q.C.; Xu, H. $Ba_{0.5}Sr_{0.5}Co_{0.8}Fe_{0.2}O_{3-\delta}$-based dual-gradient cathodes for solid oxide fuel cells. *Ceram. Int.* **2018**, *44*, 4516–4519. [CrossRef]
5. Ding, H.; Xue, X. $PrBa_{0.5}Sr_{0.5}Co_2O_{5+\delta}$ layered perovskite cathode for intermediate temperature solid oxide fuel cells. *Electrochim. Acta* **2010**, *55*, 3812–3816. [CrossRef]
6. Stambouli, A.; Traversa, E. Solid oxide fuel cells (SOFCs): A review of an environmentally clean and efficient source of energy. *Renew. Sustain. Energy Rev.* **2002**, *6*, 433–455. [CrossRef]
7. Chen, Y.; Wang, F.; Chen, D.; Dong, F.; Park, H.J.; Kwak, C.; Shao, Z. Role of silver current collector on the operational stability of selected cobalt-containing oxide electrodes for oxygen reduction reaction. *J. Power Sources* **2012**, *210*, 146–153. [CrossRef]
8. Zhou, W.; Shao, Z.; Ran, R.; Jin, W.; Xu, N. A novel efficient oxide electrode for electrocatalytic oxygen reduction at 400–600 °C. *Chem. Commun.* **2008**, *44*, 5791–5793. [CrossRef] [PubMed]
9. Yu, X.; Long, W.; Jin, F.; He, T. Cobalt-free perovskite cathode materials $SrFe_{1-x}Ti_xO_{3-\delta}$ and performance optimization for intermediate-temperature solid oxide fuel cells. *Electrochim. Acta* **2014**, *123*, 426–434. [CrossRef]
10. Rehman, A.U.; Li, M.; Knibbe, R.; Khan, M.S.; Peterson, V.K.; Brand, H.E.A.; Li, Z.; Zhou, W.; Zhu, Z. Enhancing Oxygen Reduction Reaction Activity and CO_2 Tolerance of Cathode for Low-Temperature Solid Oxide Fuel Cells by in Situ Formation of Carbonates. *ACS Appl. Mater. Interfaces* **2019**, *11*, 26909–26919. [CrossRef]
11. Liou, Y.C.; Chen, Y.R. Synthesis and microstructure of $(LaSr)MnO_3$ and $(LaSr)FeO_3$ ceramics by a reaction-sintering process. *Ceram. Int.* **2008**, *34*, 273–278.
12. Dong, F.; Chen, D.; Chen, Y.; Zhao, Q.; Shao, Z. La-doped $BaFeO_{3-\delta}$ perovskite as a cobalt-free oxygen reduction electrode for solid oxide fuel cells with oxygen-ion conducting electrolyte. *J. Mater. Chem.* **2012**, *22*, 15071–15079. [CrossRef]
13. Baiyee, Z.M.; Chen, C.; Ciucci, F. A DFT+U study of A-site and B-site substitution in $BaFeO_{3-\delta}$. *Phys. Chem. Chem. Phys.* **2015**, *17*, 23511–23520. [CrossRef]
14. Wang, J.; Saccoccio, M.; Chen, D.J.; Gao, Y.; Chen, C.; Ciucci, F. The effect of A-site and B-site substitution on $BaFeO_{3-\delta}$: An investigation as a cathode material for intermediate-temperature solid oxide fuel cells. *J. Power Sources* **2015**, *297*, 511–518. [CrossRef]
15. Dong, F.; Ni, M.; He, W.; Chen, Y.; Yang, G.; Chen, D.; Shao, Z. An efficient electrocatalyst as cathode material for solid oxide fuel cells: $BaFe_{0.95}Sn_{0.05}O_{3-\delta}$. *J. Power Sources* **2016**, *326*, 459–465. [CrossRef]
16. Dong, F.; Chen, Y.; Ran, R.; Chen, D.; Tadé, M.O.; Liu, S.; Shao, Z. $BaNb_{0.05}Fe_{0.95}O_{3-\delta}$ as a new oxygen reduction electrocatalyst for intermediate temperature solid oxide fuel cells. *J. Mater. Chem. A* **2013**, *1*, 9781–9791. [CrossRef]
17. Wang, J.; Lam, K.Y.; Saccoccio, M.; Gao, Y.; Chen, D.; Ciucci, F. Ca and in co-doped $BaFeO_{3-\delta}$ as a cobalt-free cathode material for intermediate-temperature solid oxide fuel cells. *J. Power Sources* **2016**, *324*, 224–232. [CrossRef]
18. Gao, L.; Zhu, M.Z.; Xia, T.; Li, Q.; Li, T.S.; Zhao, H. Ni-doped $BaFeO_{3-\delta}$ perovskite oxide as highly active cathode electrocatalyst for intermediate-temperature solid oxide fuel cells. *Electrochim. Acta* **2018**, *289*, 428–436.
19. Lu, Y.; Zhao, H.L.; Cheng, X.; Jia, Y.B.; Du, X.F.; Fang, M.Y.; Du, Z.H.; Zheng, K.; Świerczek, K. Investigation of In-doped $BaFeO_{3-\delta}$ perovskite-type oxygen permeable membranes. *J. Mater. Chem. A* **2015**, *3*, 6202–6214. [CrossRef]
20. Liu, X.T.; Zhao, H.L.; Yang, J.Y.; Li, Y.; Chen, T.; Lu, X.G. Lattice characteristics, structure stability and oxygen permeability of $BaFe_{1-x}Y_xO_{3-\delta}$ ceramic membranes. *J. Membr. Sci.* **2011**, *383*, 235–240.

21. Lu, Y.; Zhao, H.L.; Chang, X.W.; Du, X.F.; Li, K.; Ma, Y.H.; Yi, S.; Du, Z.H.; Zheng, K.; Świerczek, K. Novel cobalt-free $BaFe_{1-x}Gd_xO_{3-\delta}$ perovskite membranes for oxygen separation. *J. Mater. Chem. A* **2016**, *4*, 10454–10466.

22. Gao, L.; Li, Q.; Sun, L.P.; Zhang, X.F.; Huo, L.H.; Zhao, H.; Grenier, J. A novel family of Nb-doped $Bi_{0.5}Sr_{0.5}FeO_{3-\delta}$ perovskite as cathode material for intermediate-temperature solid oxide fuel cells. *J. Power Sources* **2017**, *371*, 86–95. [CrossRef]

23. Lu, F.; Xia, T.; Li, Q.; Wang, J.; Huo, L.; Zhao, H. Heterostructured simple perovskite nanorod-decorated double perovskite cathode for solid oxide fuel cells: Highly catalytic activity, stability and CO_2-durability for oxygen reduction reaction. *Appl. Catal. B Environ.* **2019**, *249*, 19–31. [CrossRef]

24. Zhu, M.; Cai, Z.; Xia, T.; Li, Q.; Huo, L.; Zhao, H. Cobalt-free perovskite $BaFe_{0.85}Cu_{0.15}O_{3-\delta}$ cathode material for intermediate-temperature solid oxide fuel cells. *Int. J. Hydrog. Energy* **2016**, *41*, 4784–4791. [CrossRef]

25. Xia, W.W.; Li, Q.; Sun, L.P.; Huo, L.H.; Zhao, H. Enhanced electrochemical performance and CO_2 tolerance of $Ba_{0.95}La_{0.05}Fe_{0.85}Cu_{0.15}O_{3-\delta}$ as Fe-based cathode electrocatalyst for solid oxide fuel cells. *J. Eur. Ceram. Soc.* **2020**, *40*, 1967–1974. [CrossRef]

26. Cai, H.D.; Xu, J.S.; Wu, M.; Long, W.; Zhang, L.; Song, Z.Y.; Zhang, L.L. A novel cobalt-free $La_{0.5}Ba_{0.5}Fe_{0.95}Mo_{0.05}O_{3-\delta}$ electrode for symmetric solid oxide fuel cell. *J. Eur. Ceram. Soc.* **2020**, *40*, 4361–4365. [CrossRef]

27. Zapata-Ramírez, V.; Mather, G.C.; Azcondo, M.T.; Amador, U.; Pérez-Coll, D. Electrical and electrochemical properties of the $Sr(Fe,Co,Mo)O_{3-\delta}$ system as air electrode for reversible solid oxide cells. *J. Power Sources* **2019**, *437*, 226895. [CrossRef]

28. Cai, W.; Guo, Y.; Zhang, T.; Guo, T.; Chen, H.; Lin, B.; Ou, X.; Liu, X. Characterization and polarization DRT analysis of a stable and highly active proton-conducting cathode. *Ceram. Int.* **2018**, *44*, 14297–14302. [CrossRef]

29. Xia, J.; Wang, C.; Wang, X.F.; Bi, L.; Zhang, Y.X. A perspective of DRT analysis for electrodes in solid oxide cells. *Electrochim. Acta* **2020**, *349*, 136328.

30. Li, S.L.; Zhang, L.K.; Xia, T.; Li, Q.; Sun, L.P.; Huo, L.H.; Zhao, H. Synergistic effect study of $EuBa_{0.98}Co_2O_{5+\delta}$-$Ce_{0.8}Sm_{0.2}O_{1.9}$ composite cathodes for intermediate-temperature solid oxide fuel cells. *J. Alloy. Comp.* **2019**, *771*, 513–521. [CrossRef]

31. Escudero, M.; Aguadero, A.; Alonso, J.; Daza, L. A kinetic study of oxygen reduction reaction on La_2NiO_4 cathodes by means of impedance spectroscopy. *J. Electroanal. Chem.* **2007**, *611*, 107–116. [CrossRef]

32. Takeda, Y.; Kanno, R.; Noda, M.; Tomida, Y.; Yamamoto, O. Cathodic Polarization Phenomena of Perovskite Oxide Electrodes with Stabilized Zirconia. *J. Electrochem. Soc.* **1987**, *134*, 2656–2661. [CrossRef]

33. Li, J.; Huo, J.; Lu, Y.; Wang, Q.; Xi, X.; Fan, Y.; Fu, X.Z.; Luo, J.L. Ca-containing $Ba_{0.95}Ca_{0.05}Co_{0.4}Fe_{0.4}Zr_{0.1}Y_{0.1}O_{3-\delta}$ cathode with high CO_2-poisoning tolerance for proton-conducting solid oxide fuel cells. *J. Power Sources* **2020**, *453*, 227909. [CrossRef]

34. Bucher, E.; Egger, A.; Caraman, G.B.; Sitte, W. Stability of the SOFC cathode materials $(Ba,Sr)(Co,Fe)O_{3-\delta}$ in CO_2-containing atmospheres. *J. Electrochem. Soc.* **2008**, *155*, B1218–B1224. [CrossRef]

35. Gu, H.; Sunarso, J.; Yang, G.; Zhou, C.; Song, Y.; Zhang, Y.; Wang, W.; Ran, R.; Zhou, W.; Shao, Z. Turning Detrimental Effect into Benefits: Enhanced Oxygen Reduction Reaction Activity of Cobalt-Free Perovskites at Intermediate Temperature via CO_2-Induced Surface Activation. *ACS Appl. Mater. Interfaces* **2020**, *12*, 16417–16425. [CrossRef]

36. Zhang, Y.; Yang, G.; Chen, G.; Ran, R.; Zhou, W.; Shao, Z. Evaluation of the CO_2 Poisoning Effect on a Highly Active Cathode $SrSc_{0.175}Nb_{0.025}Co_{0.8}O_{3-\delta}$ in the Oxygen Reduction Reaction. *ACS Appl. Mater. Interfaces* **2016**, *8*, 3003–3011. [CrossRef]

37. Zhu, Y.; Zhou, W.; Chen, Y.; Shao, Z. An Aurivillius Oxide Based Cathode with Excellent CO_2 Tolerance for Intermediate-Temperature Solid Oxide Fuel Cells. *Angew. Chem.* **2016**, *128*, 9134–9139. [CrossRef]

38. Wang, S.F.; Yeh, C.T.; Wang, Y.R.; Hsu, Y.F. Effect of $(LaSr)(CoFeCu)O_{3-\delta}$ cathodes on the characteristics of intermediate temperature solid oxide fuel cells. *J. Power Sources* **2012**, *201*, 18–25. [CrossRef]

39. Shi, H.; Ding, Z.; Ma, G. Electrochemical Performance of Cobalt-free $Nd_{0.5}Ba_{0.5}Fe_{1-x}Ni_xO_{3-\delta}$ Cathode Materials for Intermediate Temperature Solid Oxide Fuel Cells. *Fuel Cells* **2016**, *16*, 258–262. [CrossRef]

40. Zhu, Z.; Wei, Z.; Zhao, Y.; Chen, M.; Wang, S. Properties characterization of tungsten doped strontium ferrites as cathode materials for intermediate temperature solid oxide fuel cells. *Electrochim. Acta* **2017**, *250*, 203–211. [CrossRef]

41. Song, X.Q.; Le, S.R.; Zhu, X.D.; Qin, L.; Luo, Y.; Li, Y.W.; Sun, K.N.; Chen, Y. High performance $BaFe_{1-x}Bi_xO_{3-\delta}$ as cobalt-free cathodes for intermediate temperature solid oxide fuel cell. *Int. J. Hydrog. Energy* **2017**, *42*, 15808–15817. [CrossRef]

42. Ren, R.; Wang, Z.; Meng, X.; Xu, C.; Qiao, J.; Sun, W.; Sun, K. Boosting the Electrochemical Performance of Fe-Based Layered Double Perovskite Cathodes by Zn^{2+} Doping for Solid Oxide Fuel Cells. *ACS Appl. Mater. Interfaces* **2020**, *12*, 23959–23967. [CrossRef] [PubMed]

43. Xie, D.; Guo, W.; Guo, R.; Liu, Z.; Sun, D.; Meng, L.; Zheng, M.; Wang, B. Synthesis and electrochemical properties of $BaFe_{1-x}Cu_xO_{3-\delta}$ perovskite oxide for IT-SOFC cathode. *Fuel Cells* **2016**, *16*, 829–838. [CrossRef]

44. Ni, W.; Zhu, T.; Chen, X.; Zhong, Q.; Ma, W. Stable, efficient and cost-competitive Ni-substituted Sr $(Ti,Fe)O_3$ cathode for solid oxide fuel cell: Effect of A-site deficiency. *J. Power Sources* **2020**, *451*, 227762. [CrossRef]

Publisher's Note: MDPI stays neutral with regard to jurisdictional claims in published maps and institutional affiliations.

© 2020 by the authors. Licensee MDPI, Basel, Switzerland. This article is an open access article distributed under the terms and conditions of the Creative Commons Attribution (CC BY) license (http://creativecommons.org/licenses/by/4.0/).

Article

Exploring the Effect of NiO Addition to $La_{0.99}Ca_{0.01}NbO_4$ Proton-Conducting Ceramic Oxides

Kaili Yuan [1,†], Xuehua Liu [1,*,†] and Lei Bi [2]

1   Institute of Materials for Energy and Environment, College of Materials Science and Engineering, Qingdao University, Ningxia Road No. 308, Qingdao 266071, China; 2018025338@qdu.edu.cn
2   School of Resource Environment and Safety Engineering, University of South China, Hengyang 421001, China; lei.bi@usc.edu.cn or bilei81@gmail.com
*   Correspondence: liuxuehua@qdu.edu.cn
†   These authors contributed equally to this paper.

Abstract: To improve the performance and overcome the processing difficulties of $La_{0.99}Ca_{0.01}NbO_4$ proton-conducting ceramic oxide, external and internal strategies were used, respectively, to modify $La_{0.99}Ca_{0.01}NbO_4$ with NiO. The external strategy refers to the use of the NiO as a sintering aid. The NiO was added to the synthesized $La_{0.99}Ca_{0.01}NbO_4$ powder as a secondary phase, which is the traditional way of using the NiO sintering aid. The internal strategy refers to the use of NiO as a dopant for the $La_{0.99}Ca_{0.01}NbO_4$. Both strategies improve the sinterability and conductivity, but the effect of internal doping is more significant in enhancing both grain growth and conductivity, making it more desirable for practical applications. Subsequently, the influences of different concentrations of NiO were compared to explore the optimal ratio of the NiO as the dopant. It was found that the sample with 1 or 2 wt.% NiO had similar performance, while with 5 wt.%, NiO doping content hampered the grain growth. In addition, the inhomogeneous distribution of the element in the high-NiO content sample was found to be detrimental to the electrochemical performance, suggesting that the moderate doping strategy is suitable for $La_{0.99}Ca_{0.01}NbO_4$ proton-conducting electrolyte with improved performance. Furthermore, first-principle calculations indicate the origin of the enhanced performance of the internally modified sample, as it lowers both oxygen formation energy and hydration energy compared with the un-modified one, facilitating proton migration.

Keywords: proton-conducting oxide; LaNbO4; NiO; sintering; theoretical calculations

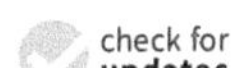

Citation: Yuan, K.; Liu, X.; Bi, L. Exploring the Effect of NiO Addition to $La_{0.99}Ca_{0.01}NbO_4$ Proton-Conducting Ceramic Oxides. *Coatings* **2021**, *11*, 562. https://doi.org/10.3390/coatings11050562

Academic Editor: Narottam P. Bansal

Received: 6 March 2021
Accepted: 5 May 2021
Published: 11 May 2021

Publisher's Note: MDPI stays neutral with regard to jurisdictional claims in published maps and institutional affiliations.

Copyright: © 2021 by the authors. Licensee MDPI, Basel, Switzerland. This article is an open access article distributed under the terms and conditions of the Creative Commons Attribution (CC BY) license (https://creativecommons.org/licenses/by/4.0/).

1. Introduction

Solid oxide fuel cells (SOFCs) can now be divided into two types depending on the properties of the electrolyte, including oxygen–ion conducting solid oxide fuel cells [1] and proton-conducting solid oxide fuel cells [2]. Classical SOFCs (oxygen-ion conducting electrolyte) require high working temperatures (>700 °C), which results in many problems, such as electrode sintering, diffusion at the interface, and difficulty in the preparation of seals and interconnection [3]. At the same time, classical SOFCs produce water at the anode side (fuel side), which would dilute the fuel and reduce fuel efficiency. Furthermore, H_2O is likely to oxidize the anode under high loads. In comparison, proton-conducting SOFCs would permit a reduction in working temperatures due to the lower activation energy for proton migration than that for oxygen-ions [4,5].

Meanwhile, water is formed at the cathode side, so the fuel is not diluted, and the anode avoids the danger of being oxidized even at high current conditions [6]. Consequently, proton-conducting SOFCs are currently a popular topic in the field of SOFCs [7,8].

The current state-of-the-art materials are acceptor-doped $BaCeO_3$ and $BaZrO_3$ [9]. Although $BaCeO_3$-based material has excellent protonic conductivity in the order of 10^{-2} $S \cdot cm^{-1}$ (600 °C), it tends to be vulnerable in the acidic gas environment, such as CO_2 and H_2O [10]. In contrast, $BaZrO_3$ has excellent chemical stability but requires a high sintering temperature and exhibits high grain boundary resistance [11]. These materials both have their own merits and limitations, so either their properties must be enhanced or new electrolytes must be found. Among all the new types of proton-conducting oxides, $LaNbO_4$ (LNO), which is reported to show pure protonic conductivity below 800 °C under wet reducing conditions, is proposed. Without involving the Ba element, $LaNbO_4$ has excellent chemical stability over the whole testing temperature range, providing an advantage in practical application. Although the motivation of developing $LaNbO_4$-based proton conductors is to eradicate Ba as the main element for improved chemical stability, the relatively low conductivity of doped $LaNbO_4$ hinders its applications. It has been found that the solubility limit of the conventional dopant in $LaNbO_4$ at the La site is less than 1% [12]. The highest proton conductivity was achieved for Ca-doped LNO, reaching approximately 10^{-3} S·cm^{-1} at 800 °C [13]. In spite of the low conductivity, several works have indicated that the LNO-based cells could reach some fuel cell performance by properly tailoring the electrode and reducing the thickness of the dense electrolyte. Fuel cells with $La_{0.995}Sr_{0.005}NbO_4$ electrolyte (thickness ~30 μm) showed a maximum output power of 1.35 mW·cm^{-2} at 800 °C [14]. By tailoring the anode, fuel cells based on LNO electrolyte film deposited on LNO-NiO anodes showed a peak output power of 24 mW·cm^{-2} at 750 °C [15]. As the electrolyte material, the $LaNbO_4$ has to be sintered densely for utilization. Although the sintering ability of $LaNbO_4$-based oxides is not as low as that of $BaZrO_3$, high sintering temperatures are still needed (such as 1500 °C) for achieving the dense $LaNbO_4$ membrane [16]. It is reasonable to assume that the conductivity of the sample could be improved if the grain growth of the $LaNbO_4$ sample could be further enhanced, which reduces the grain boundary resistance [17]. The use of dopant could be a feasible approach to improve the sinterability of the samples, as it was applied before for other proton-conducting oxides [18]. However, the investigation of dopants for $LaNbO_4$ is scarce to date.

In this study, NiO was used as the dopant for $LaNbO_4$, and different strategies were used to explore the best way of using NiO as the dopant. Furthermore, the optimal NiO content was investigated, with the analysis of the electrochemical performance of the samples to explore the factors restricting the conductivity of $LaNbO_4$ with the use of an NiO dopant.

2. Materials and Methods

The $La_{0.99}Ca_{0.01}NbO_4$ (LCNO) powder was synthesized via a traditional solid-state reaction route. Briefly, the mixture of 1.61 g La_2O_3 (analytical reagent, purity > 98%), 1.33 g Nb_2O_5 (analytical reagent, purity 99.9%) and 0.01 g $CaCO_3$ (analytical reagent, purity 99.9%) was mixed via ball milling in ethanol for 24 h. The mixed sample was heated in an oven to evaporate the ethanol, and the dry powder was acquired. The powder was calcined in a furnace at 1100 °C for 5 h to obtain pure phase LCNO powder. The NiO (analytical reagent, purity 99.8%) was added to the LCNO sample in two different ways. One was the external addition, and the other was the internal doping. To obtain 1 wt.% NiO-doped LCNO by an external addition method, the pure phase LCNO powder was mixed with NiO in a mass ratio of LCNO (100):NiO (1), and the dry powder was acquired by using the same ball milling method. For comparison, La_2O_3, Nb_2O_5, $CaCO_3$, and NiO were used as starting powders to compose 1 wt.% NiO-doped LCNO powder internally. Ni partially replaced Nb in the lattice, and the amount of defect Nb_2O_5 was calculated according to the molar amount of the NiO. The powder was also calcined at 1100 °C for 5 h to obtain pure phase NiO-doped LCNO powder. The phase structures of the powders were identified by X-ray diffraction (XRD, with CuKα radiation, Ultima IV, Rigaku, Tokyo, Japan). The angle range is from 20° to 80°. The scanning rate is 3° per minute.

Both powders were pressed into pellets and sintered at 1400 °C for 5 h to densify the pellet for the conductivity tests. The morphologies of the sintered samples were observed by a field-emission scanning electron microscope (SEM, CHI760E, JEOL, Tokyo, Japan), and the elemental distribution of the sintered membranes was analyzed by energy-dispersive X-ray spectroscopy (EDS, X-Max 50, Oxford Instruments, Oxford, UK).

The conductivity tests were carried out for the dense LCNO pellets that had been sintered at 1400 °C. To test the conductivity of the samples, both sides of sintered pellets were painted with silver paste, then the pellets were heated at 800 °C for 2 h to remove organics and form the silver electrodes for electrochemical testing. The pellets were tested in a fuel cell condition to explore the conductivity of the oxides under the fuel cell testing condition, using an electrochemical workstation (Admiral Plus, Admiral Instrument, Tempe, AZ, USA). The range of experiment temperatures is from 700 to 400 °C, and the conductivity was measured in 50 °C intervals. The single cells were tested with humidified hydrogen (~3% H_2O) as the fuel, with a flowing rate of 20 mL·min^{-1} and static air as the oxidant. We used the four-probe method for the conductivity measurement, and it was measured by an impedance spectroscopy method. The fitting of the impedance was performed using the RelaxIS software (RelaxIS 3, rhd instruments GmbH & Co. KG, Darmstadt, Germany) with a model of two distributed elements composed of a constant phase element in parallel with a resistance. Theoretical calculations were carried out by using the VASP software (VASP 6.1.1, University of Vienna, Vienna, Austria), and the calculation details can be found in our previous studies [19–22].

3. Results and Discussion

3.1. Phase, Surface and Conductivity Comparisons of Different Doping LCNO

Figure 1 shows the X-ray diffraction (XRD) patterns of LCNO powder and 1 wt.% NiO-doped LCNO powders (externally and internally) after firing at 1100 °C. One can see that both doped powders are pure phases without any detectable secondary phases. The pure phase 1 wt.% NiO internally doped powder suggests that the material still has the original structure, and the doping of NiO does not change the material structure. However, it is interesting that 1 wt.% NiO externally doped powder is also free of any impurities. During powder synthesis, the NiO source was added externally to obtain a composite of LCNO and 1 wt.% NiO, which means that the NiO peak should be shown in the XRD pattern. The absence of the NiO peak is probably due to the very low amount of NiO used in the current case, which leads to undefined NiO peaks in the XRD pattern. By comparing the LCNO XRD with the standard LNO PDF card (22–1125), we find that this phase of the synthesized powders agrees well with the standard PDF card (22–1125) of LNO, suggesting that they are compounds instead of individual oxides. The peaks at 35, 38 and 57 reflect (200), (−211)/(112) and (−321) planes of the LNO, respectively. We can see there are no obvious extra peaks, suggesting that the materials are pure phase. The shapes of the peaks show some differences, which may result from the incorporation of NiO.

Figure 2 shows the surface of the LCNO and 1 wt.% NiO-doped LCNO (externally and internally) electrolyte membranes after sintering at 1400 °C for 5 h. One can see that all three of these pellets are dense and without noticeable pores, suggesting that 1400 °C is a sufficient temperature to densify the LCNO samples. However, despite the high density obtained for the NiO-free sample, its grain size is obviously smaller than that of the NiO-modified samples, indicating that the use of NiO as the dopant is helpful for the grain growth of LCNO, regardless of if the NiO is added externally or internally. Further comparing the external and internal strategies, the sample with the NiO added internally has a larger grain size compared with the sample with the NiO added externally. By using the line interception procedure, average grain sizes of LCNO and 1 wt.% NiO-doped LCNO (externally and internally) were calculated as 0.76, 1.26 and 1.61 μm, respectively. It is understood that the large grain size could reduce the volume of the grain boundaries [23] and thus decrease the grain boundary resistance and increase the total conductivity. Therefore, it is expected that the internally doped LCNO sample should have better conductivity.

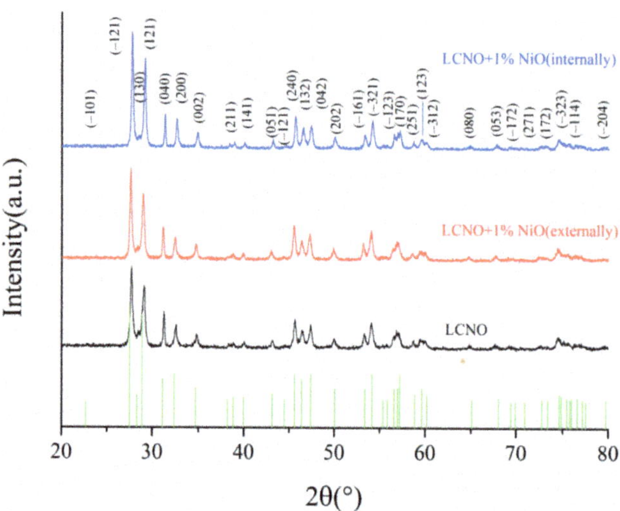

Figure 1. XRD patterns for LCNO powder and 1 wt.% NiO-doped LCNO powders (externally and internally) fired at 1100 °C.

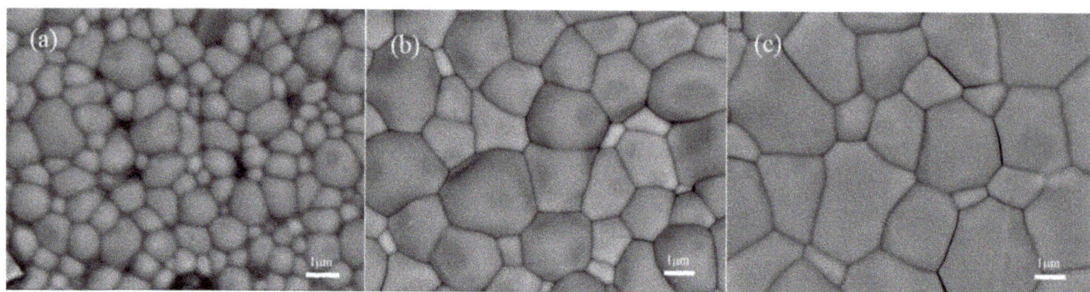

Figure 2. SEM images (shown in BSE) for the surface of (**a**) LCNO, (**b**) 1 wt.% NiO externally doped LCNO and (**c**) 1 wt.% NiO internally doped LCNO.

As shown in Figure 3, the electrical conductivity of internal doping over the entire temperature range is significantly higher than that of external doping and the LCNO pellet without NiO modification. The conductivity of the undoped LCNO, externally doped LCNO and internally doped LCNO is 0.45×10^{-3}, 0.98×10^{-3} and 1.16×10^{-3} S·cm^{-1} at 700 °C, respectively. This result indicates that the strategy of using NiO as an internal dopant is superior to the traditional method of adding NiO externally.

It is evident that the strategy of using NiO internally can further improve the conductivity of LCNO compared with the traditional way of using NiO as the external sintering aid. Therefore, further explorations of the optimal content of NiO for LCNO were carried out. Figure 4a shows the X-ray diffraction (XRD) patterns of 1, 2 and 5 wt.% NiO-doped LCNO powders by an internal addition method after firing at 1100 °C. The pure phase was formed in 1 wt.% doping sample, and the XRD pattern for the 2 wt.% doping powder is almost the same as the 1 wt.% doping sample, which indicates that the sample also possesses the original structure. However, it is interesting that the 5 wt.% NiO-doped LCNO powder has extra peaks compared to 1 wt.% doping, suggesting this powder was not a pure phase and there were some other impurities. We hypothesize that the excessive peaks come from the NiO as the concentration of 5 wt.% NiO was too high to fully incorporate into the LCNO lattice. To confirm this, we compared the XRD pattern of pure phase powder with that of the 5 wt.% doping sample and found that all extra peaks matched well with the peaks for NiO, and the result is shown in Figure 4b. Undoubtedly, 5 wt.% NiO-doped LCNO powder contains NiO as the impurity, while 1 and 2 wt.% NiO-doped LCNO powders are pure phase.

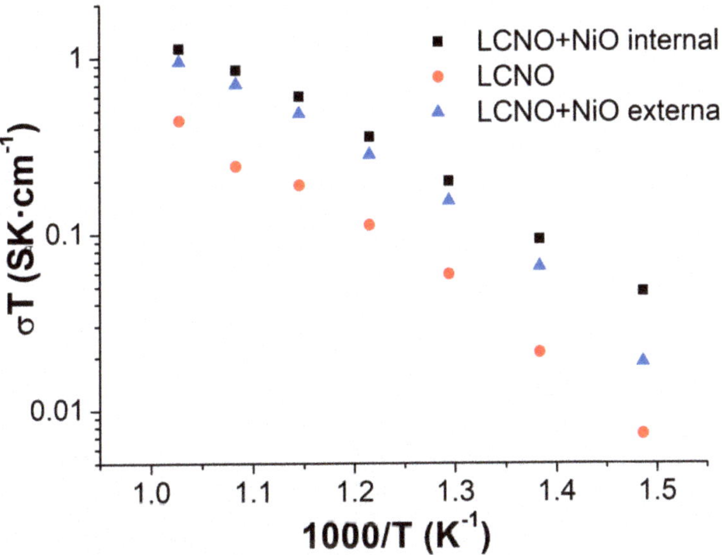

Figure 3. The conductivities of the LCNO pellet without NiO addition, and LCNO with 1 wt.% NiO addition externally and internally.

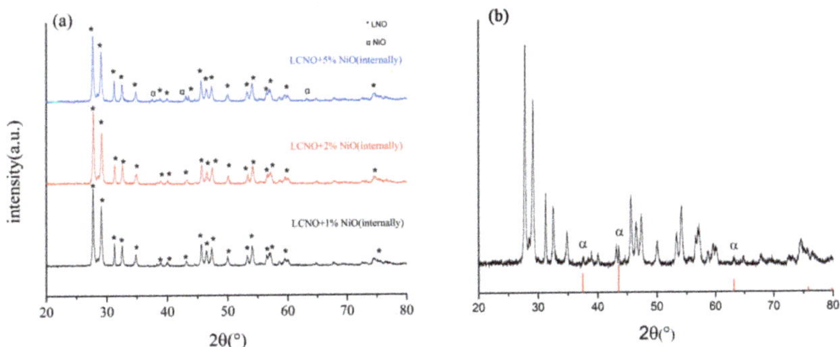

Figure 4. (**a**) XRD patterns for 1, 2 and 5 wt.% NiO-doped LCNO powders by an internal addition method fired at 1100 °C; (**b**) labels of excessive peaks from 5 wt.% doping XRD patterns.

Figure 5 shows the surface of 1, 2 and 5 wt.% NiO-doped LCNO electrolyte membranes after sintering at 1400 °C for 5 h. It can be observed that all three of these pellets are dense. However, there are some differences in the grain size of the pellets. The grain size of the 1 and 2 wt.% doped samples is similar, while that of the 5 wt.% doped sample is relatively smaller, which may be due to the extra NiO that inhibits the grain growth. By using the line interception procedure, the average grain sizes of 1, 2 and 5 wt.% NiO-doped LCNO were calculated as 1.61, 1.53 and 1.32 μm, respectively. There is a grain with a deep color in Figure 5c, which indicates that there is a Ni element accumulation in this place. We assume that this is due to the uneven distribution of NiO, which is confirmed in the SEM-EDS results.

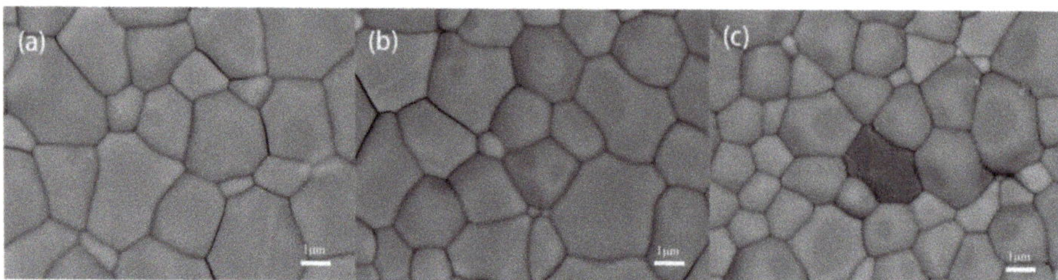

Figure 5. SEM images (shown in BSE) for the surface of (**a**) 1 wt.% NiO internally doped LCNO, (**b**) 2 wt.% NiO internally doped LCNO and (**c**) 5 wt.% NiO internally doped LCNO.

The elemental distribution of the samples was further analyzed by SEM-EDS. Figure 6 shows the SEM-EDS results for 1, 2 and 5 wt.% NiO-doped LCNO electrolyte membranes. The elemental analysis indicates that a relatively homogeneous Ni distribution is presented for 1 and 2 wt.% doped pellets. In contrast, it can be observed that the Ni element is unevenly distributed for 5 wt.% NiO-doped electrolyte, and an obvious accumulation of Ni element can be detected that agrees well with the XRD analysis. It would be reasonable to assume that the accumulation of NiO, as well as the smaller grain size, could deteriorate the conductivity of the sample.

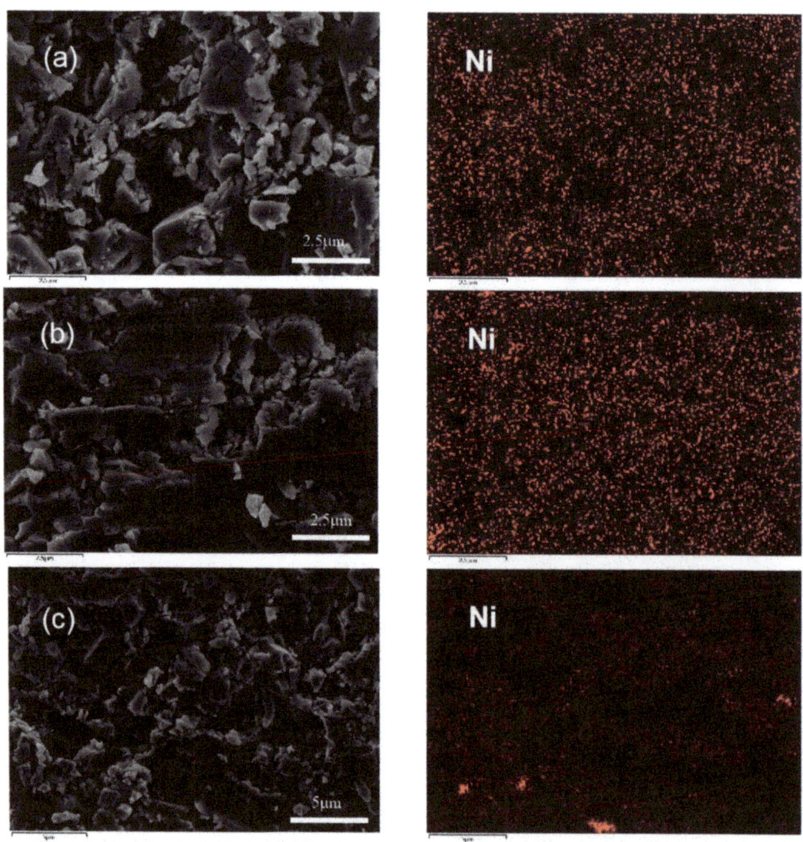

Figure 6. SEM-EDS mapping results for (**a**) 1 wt.% NiO internally doped LCNO, (**b**) 2 wt.% NiO internally doped LCNO and (**c**) 5 wt.% NiO internally doped LCNO.

Figure 7 shows the conductivity of 1, 2 and 5 wt.% NiO-doped LCNO electrolyte. One can see that the performances of 1 and 2 wt.% doping are relatively favorable, but the effect of 5 wt.% NiO on performance improvement is small. As shown in Figure 5, the grain sizes of 1 and 2 wt.% internal doping are slightly larger than that of 5 wt.%. Evidently, it could be easier for protons to transfer into samples with larger grain sizes, which reduce the grain boundary resistance. Additionally, compared to the first two samples, the element distribution of 5 wt.% doping is more uneven. Therefore, it can be concluded that 1 and 2 wt.% doping samples showed larger grain sizes and a more homogeneous distribution of elements. This may be why the performances of the first two samples are better than the 5 wt.% doped one. If we examine the conductivity difference between the 1 and 2% samples, they show similar conductivity at high temperatures, but the difference in conductivity increases at low temperatures. For instance, the conductivity for the 1, 2 and 5% doped sample is 1.16×10^{-3}, 1.25×10^{-3} and 0.49×10^{-3} S·cm^{-1} at 700 °C, respectively. In contrast, the conductivity value at 400 °C is 7×10^{-5}, 3.1×10^{-5} and 1.1×10^{-5} S·cm^{-1} for the 1%, 2% and 5% doped samples, respectively. The conductivity for the 1% sample is higher than that of the traditional LCNO reported in the literature that reaches around 1×10^{-3} S·cm^{-1} at 800 °C [16]. The observed inflection point on the dependences both in Figures 3 and 7 may be a result of the phase transformation of the LCNO-based material, as reported in the literature [16]. The activation energy was calculated for the samples, indicating the activation energy of 0.62, 0.74 and 0.77 eV for 1%, 2% and 5% samples, respectively. The change in the activation energy may be related to the composition change that we elaborate on in the following section.

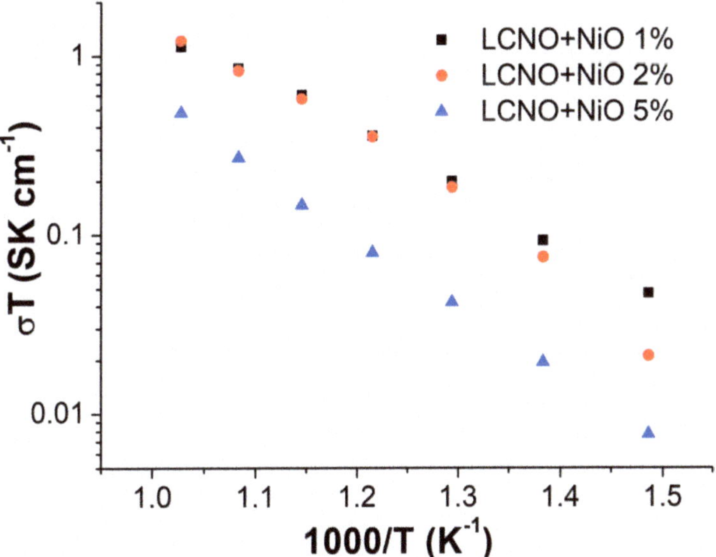

Figure 7. The conductivity of 1, 2 and 5 wt.% NiO-doped LCNO electrolyte.

The lattice parameters of pure LCNO; 1% external doping; and 1%, 2% and 5% internal doping samples were calculated as 332.884, 333.260, 332.782, 332.845, 333.167 Å^3, respectively. Rietveld refined XRD patterns are also shown in Figure 8. The lattice volume would be expected to decrease if Ni replaced Nb and was completely doped into the lattice, as the ionic radius of Ni^{2+} (60 pm) is smaller than that of Nb^{5+} (64 pm) [24,25]. When the 1% is applied with the internal method, the expected decrease in lattice volume is observed. However, the lattice volume increases from 1% to 2%, although the lattice volume of the 2% sample is still smaller than that of the undoped LCNO, implying that Ni is not fully incorporated into the lattice. The increase in lattice volume is more obvious prominent in the 5% doped sample and the sample with 1% NiO added externally (333.260 Å^3), suggesting that the limitation in lattice solution could increase the lattice volume. Although no obvious accumulation of NiO can be observed in SEM and XRD for the 2% sample, likely due to the low concentration beyond the detection of the instruments, the change in the lattice volume suggests that the solubility limit for NiO-doped LCNO is between 1 and 2%. It seems that not all the 2% NiO is incorporated into the LCNO lattice, which might be the reason for the decreased conductivity at lower temperatures and higher activation energy compared with that of the 1% doped sample.

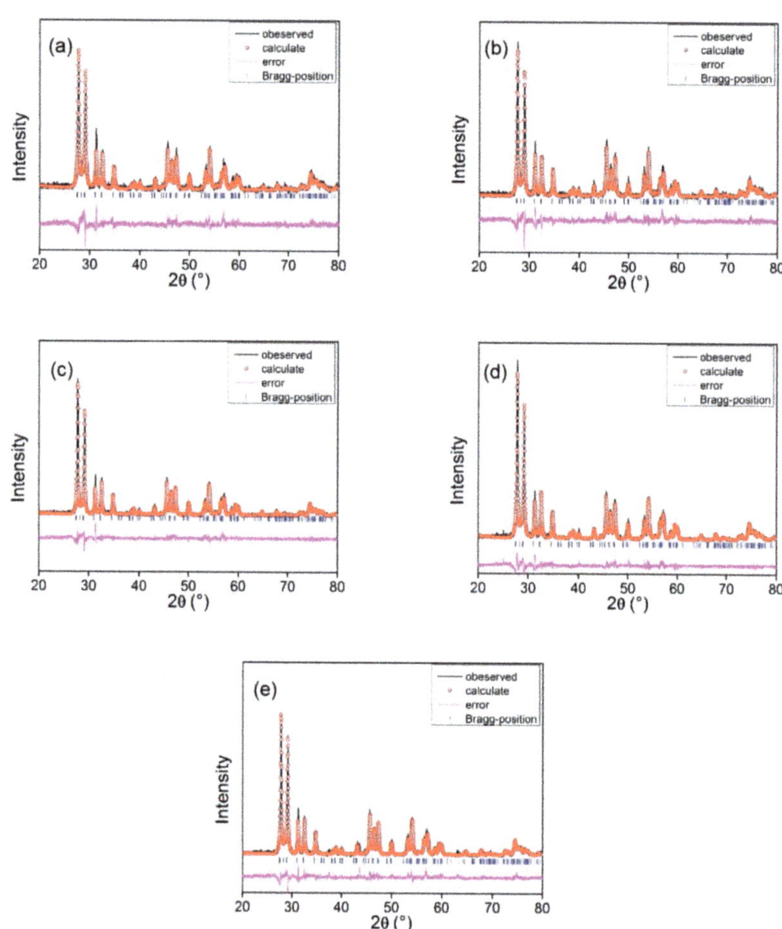

Figure 8. Rietveld refined XRD patterns of (**a**) undoped LCNO, (**b**) 1% externally doped LCNO, (**c**) 1% internally doped LCNO, (**d**) 2% internally doped LCNO and (**e**) 5% internally doped LCNO.

3.2. DFT

It is noted that the conductivity of the 1 and 2 wt.% NiO-doped LCNO shows higher conductivity than that of the LCNO pellets sintered at higher temperatures, as reported in the literature [16], implying that the tailoring of LCNO with NiO is an effective strategy to promote the protonation of LCNO. In order to prove this hypothesis, first-principle calculations were carried out. It is known that protonation happens when oxygen vacancies are created, and proton defects are formed in the wet atmosphere according to the equation $H_2O + V_O^{\bullet\bullet} + O_O^{\times} \Leftrightarrow 2OH^{\bullet}$ [26].

Therefore, the oxygen vacancy formation energy (E_{V_O}) and the hydration energy ($E_{hydration}$) were calculated for LCNO with and without NiO modification. NiO was used internally as the dopant; therefore, NiO was incorporated into the lattice by replacing Nb atoms in the calculation. Figure 9 shows the optimized configuration of the LaNbO4 with and without NiO modification by the DFT method, and one can see that the Ni atom partially occupies the Nb site for the NiO-modified sample. The oxygen vacancy formation energy (E_{V_O}) was calculated as 5.98 and 0.78 eV for the samples with and without the NiO modification, respectively. The result indicates that the introduction of NiO into the lattice can significantly lower the E_{V_O}, which is likely due to the low valence of Ni compared with the Nb. The replacement of Nb^{5+} by Ni^{2+} in the lattice could generate oxygen vacancies that, in principle, benefit the protonation (hydration) procedure. The calculated hydration energy ($E_{hydration}$) is -1.31 and -1.42 eV for the sample with and without the NiO modification, respectively. One can see that both values are negative, suggesting that hydration is thermodynamically favorable in both oxides. However, a more negative value was obtained when NiO was used, suggesting that NiO modification could have a better hydration ability than that of the NiO-free sample. Although there is no explanation as to why NiO is effective in improving the conductivity of LCNO, we can confirm that the replacement of Ni for Nb in the lattice could be beneficial for the oxygen vacancy formation as well as the hydration [27,28]. In addition, the replacement of Ni for Nb can form negative charges as Ni_{Nb}''' in the lattice, which neutralizes the positive charge at the core of the space charge layer, thus mitigating the space charge layer effect and improving the conductivity accordingly [29]. Therefore, the improved conductivity of the NiO-modified LCNO is expected.

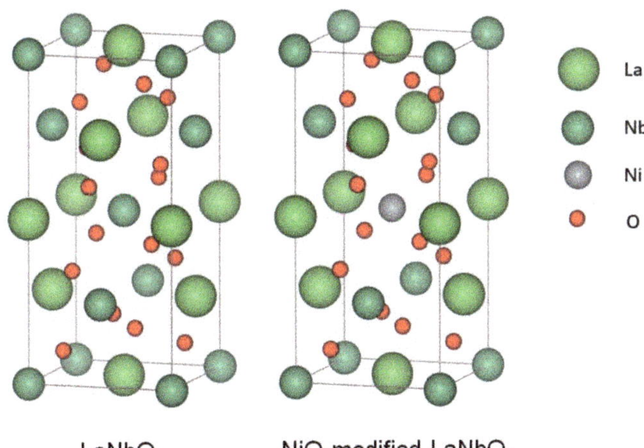

Figure 9. Configuration of LaNbO4- and NiO-modified LaNbO4 optimized by the DFT method using the VASP software.

4. Conclusions

In this study, NiO was used as a sintering aid both internally and externally to explore the influence of NiO on the performance of the $La_{0.99}Ca_{0.01}NbO_4$ (LCNO) proton-conducting oxide. NiO has a significant impact on the grain growth of LCNO, regardless of if it is used internally or externally. However, the internal doping strategy shows an advantage in both conductivity and grain growth over that of the external strategy. Furthermore, the doping concentration has an evident impact on the performance of LCNO even within the internally doped samples. The sample shows good conductivity when the doping concentration is low, and the conductivity decreases with the high NiO-doping concentration, which is due to the presence of the second phase and the accumulation of NiO. DFT calculations show that the doping of NiO into LCNO could lower the oxygen vacancy formation energy and the protonation (hydration) energy, facilitating the formation of protons and thus improving the conductivity.

Author Contributions: Investigation, K.Y. and X.L.; supervision, L.B.; writing—original draft, K.Y.; writing—review & editing, L.B. All authors have read and agreed to the published version of the manuscript.

Funding: This work was supported by the Key Research and Development Program of Shandong Province (Grant No.: 2019GGX103020).

Institutional Review Board Statement: Not applicable.

Informed Consent Statement: Not applicable.

Data Availability Statement: Data is contained within the article. The data presented in this study are available in article.

Conflicts of Interest: The authors declare no conflict of interest.

References

1. Wachsman, E.D.; Lee, K.T. Lowering the temperature of solid oxide fuel cells. *Science* **2011**, *334*, 935–939. [CrossRef] [PubMed]
2. Dai, H.; Kou, H.; Wang, H.; Bi, L. Electrochemical performance of protonic ceramic fuel cells with stable $BaZrO_3$-based electrolyte: A mini-review. *Electrochem. Commun.* **2018**, *96*, 11–15. [CrossRef]
3. Wachsman, E.; Ishihara, T.; Kilner, J. Low-temperature solid-oxide fuel cells. *MRS Bull.* **2014**, *39*, 773–779. [CrossRef]
4. Tang, H.; Jin, Z.; Wu, Y.; Liu, W.; Bi, L. Cobalt-free nanofiber cathodes for proton conducting solid oxide fuel cells. *Electrochem. Commun.* **2019**, *100*, 108–112. [CrossRef]
5. Xia, Y.; Jin, Z.; Wang, H.; Gong, Z.; Lv, H.; Peng, R.; Liu, W.; Bi, L. A novel cobalt-free cathode with triple-conduction for proton-conducting solid oxide fuel cells with unprecedented performance. *J. Mater. Chem. A* **2019**, *7*, 16136–16148. [CrossRef]
6. Duan, C.C.; Tong, J.H.; Shang, M.; Nikodemski, S.; Sanders, M.; Ricote, S.; Almansoori, A.; O'Hayre, R. Readily processed protonic ceramic fuel cells with high performance at low temperatures. *Science* **2015**, *349*, 1321–1326. [CrossRef]
7. Han, D.; Shinoda, K.; Sato, S.; Majima, M.; Uda, T. Correlation between electroconductive and structural properties of proton conductive acceptor-doped barium zirconate. *J. Mater. Chem. A* **2014**, *3*, 1243–1250. [CrossRef]
8. Tarutin, A.P.; Lyagaeva, J.G.; Medvedev, D.A.; Bi, L.; Yaremchenko, A.A. Recent advances in layered $Ln_2NiO_{4+\delta}$ nickelates: Fundamentals and prospects of their applications in protonic ceramic fuel and electrolysis cells. *J. Mater. Chem. A* **2021**, *9*, 154–195. [CrossRef]
9. Bi, L.; Boulfrad, S.; Traversa, E. Steam electrolysis by solid oxide electrolysis cells (SOECs) with proton-conducting oxides. *Chem. Soc. Rev.* **2014**, *43*, 8255–8270. [CrossRef]
10. Medvedev, D.A.; Lyagaeva, J.G.; Gorbova, E.V.; Demin, A.K.; Tsiakaras, P. Advanced materials for SOFC application: Strategies for the development of highly conductive and stable solid oxide proton electrolytes. *Prog. Mater. Sci.* **2016**, *75*, 38–79. [CrossRef]
11. Li, J.; Wang, C.; Wang, X.; Bi, L. Sintering aids for proton-conducting oxides—A double-edged sword? A mini review. *Electrochem. Commun.* **2020**, *112*, 106672. [CrossRef]
12. Haugsrud, R.; Norby, T. High-temperature proton conductivity in acceptor-substituted rare-earth ortho-tantalates, $LnTaO_4$. *J. Am. Ceram. Soc.* **2007**, *90*, 1116–1121. [CrossRef]
13. Fjeld, H.; Kepaptsoglou, D.M.; Haugsrud, R.; Norby, T. Charge carriers in grain boundaries of 0.5% Sr-doped $LaNbO_4$. *Solid State Ion.* **2010**, *181*, 104–109. [CrossRef]
14. Magraso, A.; Fontaine, M.L.; Larring, Y.; Bredesen, R.; Syvertsen, G.E.; Lein, H.L.; Grande, T.; Huse, M.; Strandbakke, R.; Haugsrud, R.; et al. Development of proton conducting SOFCs based on $LaNbO_4$ electrolyte—status in Norway. *Fuel Cells* **2011**, *11*, 17–25. [CrossRef]
15. Bi, L.; Fabbri, E.; Traversa, E. Solid oxide fuel cells with proton-conducting $La_{0.99}Ca_{0.01}NbO_4$ electrolyte. *Electrochim. Acta* **2018**, *260*, 748–754. [CrossRef]
16. Haugsrud, R.; Norby, T. Proton conduction in rare-earth ortho-niobates and ortho-tantalates. *Nat. Mater.* **2006**, *5*, 193–196. [CrossRef]
17. Haile, S.M.; Staneff, G.; Ryu, K.H. Non-stoichiometry, grain boundary transport and chemical stability of proton conducting perovskites. *J. Mater. Sci.* **2001**, *36*, 1149–1160. [CrossRef]
18. Peng, C.; Melnik, J.; Luo, J.L.; Sanger, A.R.; Chuang, K.T. $BaZr_{0.8}Y_{0.2}O_3$-delta electrolyte with and without ZnO sintering aid: Preparation and characterization. *Solid State Ion.* **2010**, *181*, 1372–1377. [CrossRef]
19. Ma, J.; Tao, Z.; Kou, H.; Fronzi, M.; Bi, L. Evaluating the effect of Pr-doping on the performance of strontium-doped lanthanum ferrite cathodes for protonic SOFCs. *Ceram. Int.* **2020**, *46*, 4000–4005. [CrossRef]

20. Xu, X.; Wang, H.; Ma, J.; Liu, W.; Wang, X.; Fronzi, M.; Bi, L. Impressive performance of proton-conducting solid oxide fuel cells using a first-generation cathode with tailored cations. *J. Mater. Chem. A* **2019**, *7*, 18792–18798. [CrossRef]
21. Xu, X.; Wang, H.; Fronzi, M.; Wang, X.; Bi, L.; Traversa, E. Tailoring cations in a perovskite cathode for proton-conducting solid oxide fuel cells with high performance. *J. Mater. Chem. A* **2019**, *7*, 20624–20632. [CrossRef]
22. Ji, Q.Q.; Xu, X.; Liu, X.H.; Bi, L. Improvement of the catalytic properties of porous lanthanum manganite for the oxygen reduction reaction by partial substitution of strontium for lanthanum. *Electrochem. Commun.* **2021**, *124*, 1873–1902. [CrossRef]
23. Fabbri, E.; Bi, L.; Tanaka, H.; Pergolesi, D.; Traversa, E. Chemically stable Pr and Y Co-doped barium zirconate electrolytes with high proton conductivity for intermediate-temperature solid oxide fuel cells. *Adv. Funct. Mater.* **2010**, *21*, 158–166. [CrossRef]
24. Zhang, C.; Wang, Y.; Li, G.; Chen, L.; Zhang, Q.; Wang, D.; Li, X.; Wang, Z. Tuning smaller Co_3O_4 nanoparticles onto HZSM-5 zeolite via complexing agents for boosting toluene oxidation performance. *Appl. Surf. Sci.* **2020**, *532*, 147320. [CrossRef]
25. Zeng, K.; Wang, Y.; Huang, C.; Liu, H.; Liu, X.; Wang, Z.; Yu, J.; Zhang, C. Catalytic combustion of propane over $MnNbO_x$ composite oxides: The promotional role of niobium. *Ind. Eng. Chem. Res.* **2021**, *60*, 6111–6120. [CrossRef]
26. Kreuer, K. Proton-conducting oxides. *Annu. Rev. Mater. Res.* **2003**, *33*, 333–359. [CrossRef]
27. Xu, Y.S.; Liu, X.H.; Cao, N.; Xu, X.; Bi, L. Defect engineering for electrocatalytic nitrogen reduction reaction at ambient conditions. *Sustain. Mater. Technol.* **2021**, *27*, e00229.
28. Xu, Y.S.; Xu, X.; Cao, N.; Wang, X.F.; Liu, X.H.; Fronzi, M.; Bi, L. Perovskite ceramic oxide as an efficient electrocatalyst for nitrogen fixation. *Int. J. Hydrogen. Energ.* **2021**, *46*, 10293–10302. [CrossRef]
29. Shirpour, M.; Rahmati, B.; Sigle, W.; van Aken, P.A.; Merkle, R.; Maier, J. Dopant segregation and space charge effects in proton-conducting $BaZrO_3$ perovskites. *J. Phys. Chem. C* **2012**, *116*, 2453–2461. [CrossRef]

Article

Enhancement Research on Piezoelectric Performance of Electrospun PVDF Fiber Membranes with Inorganic Reinforced Materials

Chong Li [1,*], Haoyu Wang [2,*], Xiao Yan [3], Hanxige Chen [4], Yudong Fu [2] and Qinhua Meng [2]

1 Center for Engineering Training, Harbin Engineering University, Harbin 150001, China
2 College of Materials Science and Chemical Engineering, Harbin Engineering University, Harbin 150001, China; fuyudong@hrbeu.edu.cn (Y.F.); qinhua97@foxmail.com (Q.M.)
3 Shandong Institute of Space Electronic Technology, Yantai 264670, China; 13351987373@163.com
4 Heilongjiang Zhongmin Measurement and Control Instrument Engineering Co., Ltd., Harbin 150001, China; m18360823175@163.com
* Correspondence: lichong@hrbeu.edu.cn (C.L.); haoyuwang@hrbeu.edu.cn (H.W.)

Abstract: The electrospun PVDF fiber membranes with the characteristics of light weight, strong signal and measurability, have been widely applied in the fields of environment, energy sensors and biomedical treatment. Due to the weakness of the piezoelectric and service properties, the conventional PVDF fiber membranes cannot meet the operating requirements. Based on the obtained optimal technological parameter of electrospun pure PVDF fiber membranes (P-PVDF) in the previous experiment (unpublished), three inorganic reinforced substances ($AgNO_3$, $FeCl_3 \cdot 6H_2O$, nanographene) were respectively used to dope and modify PVDF to prepare composite fiber membranes with the better piezoelectric performance. The morphology and crystal structure of the hybrid fiber membranes were observed and detected by scanning electron microscopy and X-ray diffraction, respectively. The results showed that the dopant could effectively promote the formation of β-phase, which can enhance the piezoelectric performance. The mechanical properties test and piezoelectric performance test exhibited that the static flexural strength, the elastic modulus, and the piezoelectric performance were improved with the addition of dopant. In addition, the influence on the addition of dopant and the doping modification mechanism were discussed. Finally, the conclusions showed that the minimum average diameter was obtained with the 0.3 wt% addition of $AgNO_3$; the piezoelectric performance reached the strongest with the 0.8 wt% addition of $FeCl_3 \cdot 6H_2O$; the mechanical properties were best with the 1.0 wt% addition of nanographene.

Keywords: electrospinning; nanomaterials; PVDF; piezoelectric performance; surface engineering

Citation: Li, C.; Wang, H.; Yan, X.; Chen, H.; Fu, Y.; Meng, Q. Enhancement Research on Piezoelectric Performance of Electrospun PVDF Fiber Membranes with Inorganic Reinforced Materials. *Coatings* **2021**, *11*, 1495. https://doi.org/10.3390/coatings11121495

Academic Editor: Günter Motz

Received: 30 October 2021
Accepted: 3 December 2021
Published: 5 December 2021

Publisher's Note: MDPI stays neutral with regard to jurisdictional claims in published maps and institutional affiliations.

Copyright: © 2021 by the authors. Licensee MDPI, Basel, Switzerland. This article is an open access article distributed under the terms and conditions of the Creative Commons Attribution (CC BY) license (https://creativecommons.org/licenses/by/4.0/).

1. Introduction

With the aggravation of the energy crisis and environmental problems, scientists have diverted their attention to safer and cleaner energy resources. The most important one is the conversion of mechanical energy to electrical energy. As one of the materials that are able to convert energy, piezoelectric materials have attracted many interests and have been used widely in various applications such as sensors, actuators, nanoelectronics, and energy harvesting [1–4]. Among them, PVDF and its co-polymer are becoming the focus of research regarding its low cost, high flexibility, quick response, and piezoelectric. These outstanding properties can be adopted as energy harvesters [5,6] and sensors [7–11].

It is known that PVDF has five different crystallite polymorphs (α, β, γ, δ, and ε) which can be transformed to each other under particular conditions [12]. The β-phase in these is the critical electroactive component of PVDF to exhibit the ferroelectric properties and piezoelectric properties [13]. However, what can be obtained by the traditional preparation process of PVDF membranes is the nonpolar α-phase [12]. The β-phase of obtained PVDF membranes should always be induced by some post-processing treatments, such as mechanical stretching, annealing, and high voltage poling [13–16]. Compared with the complex traditional preparation process of PVDF piezoelectric membranes, the electrospinning process, including the ejection of the PVDF solution, volatilization of the solvent, stretching of the electrospinning streams, and high voltage polarization, could be completed in one step and the β-phase content of the prepared PVDF membranes were improved with the function of high voltage and stretching, which exhibit the better piezoelectric properties [17]. During the previous study, the pure PVDF membranes were fabricated with the optimized electrospinning technological parameters (the ratio of solvent, the concentration of PVDF solution, the voltage of electrospinning, the injection speed PVDF solution, and the rotate speed of collector). However, the pure PVDF fiber membrane cannot meet the operating requirements because of the weakness of the piezo-electric and service properties. Meanwhile, the inorganic dopant materials into PVDF is a simple and effective approach to enhance the piezoelectric properties of PVDF [18–20]. Therefore, it is quite crucial to prepare the inorganic-PVDF fiber membranes with better piezoelectric properties.

In this study, to further improve the performance of PVDF fiber membranes, three inorganic substances ($AgNO_3$, $FeCl_3 \cdot 6H_2O$, nanographene) were applied to dope and modify PVDF to prepare organic-inorganic composite fiber membranes based on the previous study. The morphology and crystal structure of the composite fiber membranes were observed, and the mechanical properties and piezoelectric performance of that were measured. Moreover, the research provides the possibility of wider application of PVDF composite fiber membranes.

2. Experimental Details

2.1. Materials

Poly (vinylidene fluoride) (PVDF) was supplied by Sigma-Aldrich Corp (Saint Louis, MO, USA). Acetone (C_3H_6O), nitric acid (HNO_3) and N, N-dimethyl-formamide (DMF) were obtained from Tianjin Tianda Chemical Reagent Factory (Tianjin, China). Ethyl alcohol (C_2H_6O) was produced by Chinasun Specialty Products Co., Ltd. (Changshu, China). Citric acid ($C_6H_8O_7$), Ethylene Diamine Tetraacetic Acid (EDTA), carbamide ($CO(NH_2)_2$, argentum nitricum ($AgNO_3$), and ferric chloride ($FeCl_3 \cdot 6H_2O$) was purchased from Tianjin Zhiyuan Chemical Reagent Co., Ltd. (Tianjin, China). Lanthanum chloride ($LaCl_3 \cdot 6H_2O$) was provided by Jining Zhongkai New Materials Co., LTD (Jining, China). Nanographene was manufactured by Knano Graphene Technology Co., Ltd. (Xiamen, China).

2.2. Preparation of Modified Nanographene

It is well known that the nanographene is easily aggregated to unevenly dispersed in the matrix, because of its small particle size and large specific surface area. Thus, the rare earth modification solution is synthesized to modify the surface of nanographene. A certain quality of rare earth ($LaCl_3 \cdot 6H_2O$) was dissolved in the 100 mL ethanol solution. Then, according to a certain mass percent, the NH_4Cl, $CO(NH_2)_2$, EDTA, and $C_6H_8O_7$ were added into it to obtain mixed solution, as shown in Table 1. After the prepared modification solution was adequately stirred by a magnetic stirrer for 20 min and heated at 80 °C for 2 h, the PH value of that was adjusted to 4–6 by HNO_3. Letting it stand for a period of time, the clear rare earth modification solution was obtained.

Table 1. Composition of the rare earth modification solution.

Name	Molecular Formula	Mass Percent
Ethyl alcohol	C_2H_6O	90%–99%
Citric acid	$C_6H_8O_7$	0.5%–1%
Urea	$CO(NH_2)_2$	0.5%–1%
Ethylene diamine tetraacetic acid	EDTA	0.5%
Ammonium chloride	NH4Cl	0.1%–1%
Lanthanum chloride	$LaCl_3 \cdot 6H_2O$	0.5%

After 10 mg nanographene adequately dispersed in modification solution by ultrasonication for 5 h, the modified nanographene dispersion was obtained. With filter units, the modified nanographene was collected and then washed several times with ethyl alcohol and hot deionized water. Finally, the modified nanographene was obtained after drying at 80 °C for 48 h in a drying oven.

2.3. Electrospinning of Nanocomposites

Four kinds of nanocomposites, including P-PVDF fibers, PVDF/AgNO$_3$ fibers (PVDF-Ag), PVDF/FeCl$_3$·6H$_2$O fibers (PVDF-Fe) and PVDF/Nanographene fibers (PVDF-G)-, will be prepared in this experiment. Initially, the powders of reinforced materials (AgNO$_3$, FeCl$_3$·6H$_2$O and modified nanographene) were dried for 24 h at 50 °C, 50 °C and 80 °C, respectively. Meanwhile, 15% of PVDF powders were dissolved and stirred in a mixed solvent of DMF and acetone (3:2 by weight) under magnetic stirring for 2 h. Then, the solution of PVDF was heated for 3 h under the 60 °C in a water bath to guarantee adequately dissolved and homogeneous. Afterwards, the powders of AgNO$_3$ (0.1 wt%, 0.3 wt%, 0.5 wt%, 0.7 wt%) and FeCl$_3$·6 H$_2$O (0.4 wt%, 0.8 wt%, 1.2 wt%, 1.6 wt%) were added into the solution of PVDF and stirred at 60 °C for 20 min to obtain PVDF/AgNO$_3$ and PVDF/FeCl$_3$·6H$_2$O solution. The electrical conductivity of PVDF/AgNO$_3$ and PVDF/FeCl$_3$·6H$_2$O solution was accordingly measured in Figure 1. Compared with that, the PVDF/modified nanographene solution was prepared differently. The dried modified nanographene (0.5 wt%, 1.0 wt%, 1.5 wt%, 2.0 wt%) should be firstly dispersed in the DMF by ultrasonication for 2 h and then used to prepare the PVDF/modified nanographene solution. Finally, the four kinds of electrospinning solution were kept for 12 h in the syringe to exclude bubbles.

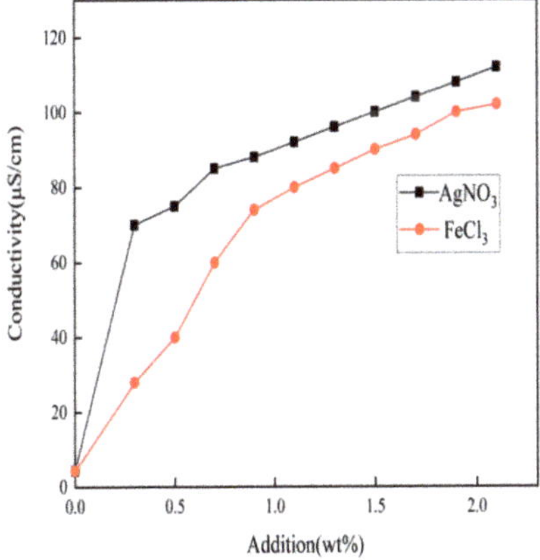

Figure 1. Electrical conductivity of PVDF/AgNO$_3$ and PVDF/FeCl$_3$·6H$_2$O solution.

The electrospinning device is shown in Figure 2. The prepared solution (P-PVDF, PVDF/AgNO$_3$, PVDF/FeCl$_3$·6H$_2$O and PVDF/nano-graphene) was respectively put into a 20 mL syringe connected to the conduit in the device. With the obtained optimal process parameters of P-PVDF, the syringe was placed in the groove with a high-voltage power supply of 22.5 kv. The roller twisted with aluminum foil was used as the collector of the electrospun fibrous membrane at the speed of 600 r/min. Electrospinning was done with an ejection rate of 1 mL/h from the syringe and 12 cm work distance between the spinneret and the collector. Finally, the prepared fibers were taken away from the roller and kept in plastic packaging bags for the coming test.

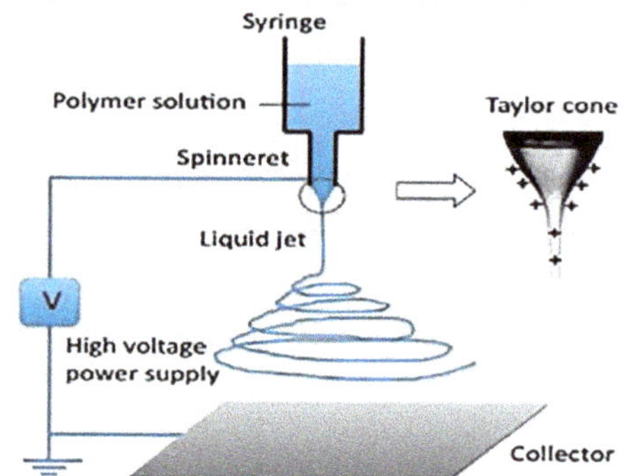

Figure 2. Electrospinning device map.

2.4. Material Characterizations

The morphology of fibers was investigated by scanning electron microscopy (SEM) (SEM, JSM-6480, Tokyo, Japan). The average diameter of fibers from the 100 randomly drawn fibers in one photo of SEM was measured by Image-J (Fiji image-J, National Institute of Health, Bethesda, Rockville, MD, USA). The crystalline phases of the samples were determined with the scan area of 10°–40° and the scan speed of 2°/min by X-ray diffraction (XRD) (Rigaku 5th miniflex, Tokyo, Japan) equipped with the Cu-Kα tube operating at 40 kv and 40 mA. The mechanical properties of the fibers were tested by a universal testing machine (UTM, WOW-50, Jinan Liangong Testing Technology Co., Ltd., Jinan, China). According to the test standard of electrospun fibers (ASTMD 638), the samples were cut with the size of 20 mm × 30 mm and loaded in a constant deformation mode at a speed of 20 mm/min by the clamping length of 20 mm fiber. The viscosity was calculated by type and rotation speed of the rotor in the rotational viscometer (NDJ79, Shanghai INESA Scientific Instrument CO., LTD, Shanghai, China). With the dried and revised probe kept for 3 min in the solution, the conductivity of the solution was obtained by the conductivity tester (DDS-11A, Shanghai Changji Geological Instrument Co., Ltd., Shanghai, China).

2.5. Piezoelectric Properties Test

The structure of the piezoelectric signal generator with the size of 25 mm × 35 mm is shown in Figure 3. Aluminum foils were covered on both sides of fiber membrane samples with conducting adhesive as the electrode of the membrane. The tape of copper foils covered on the aluminum foils was used as a conductor and staggered electrode to the delivery charge. To protect the electrode from damage and short circuit, the insulated rubber tape was used to pack the electrode patches in the outermost layer. Finally, the piezoelectric signal generator was completed to conduct a piezoelectric performance test.

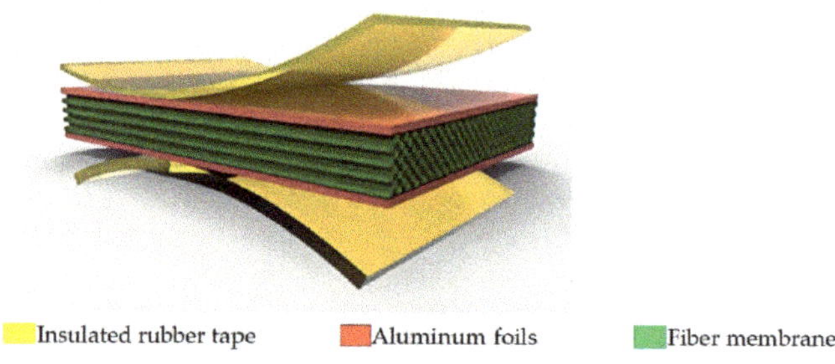

Figure 3. Piezoelectric signal generator.

As shown in Figure 4, the piezoelectric test device was composed of power plant, charge-amplifier and oscilloscope. The vibrostand as the power plant was used with a certain frequency (500 Hz) and amplitude (5 mm), as shown in Figure 5. Because of the weak electrical signal of the piezoelectric fiber membrane, the charge amplifier driven by the voltage of 5 V was used to amplify the resulting signal and restrain the influence of other interference signals. A chunk with a weight of 1 g was pressed on the piezoelectric signal generator to keep the vibration smooth. The piezoelectric signal generator fixed on the vibrostand was connected in turn to wire, charge amplifier and oscilloscope. When the piezoelectric properties test begins, the electrical signal can be harvested in the oscilloscope, and the piezoelectric test is completed.

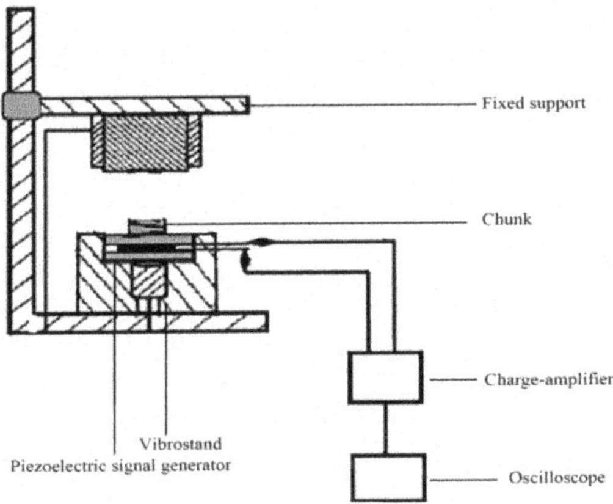

Figure 4. Piezoelectric test device.

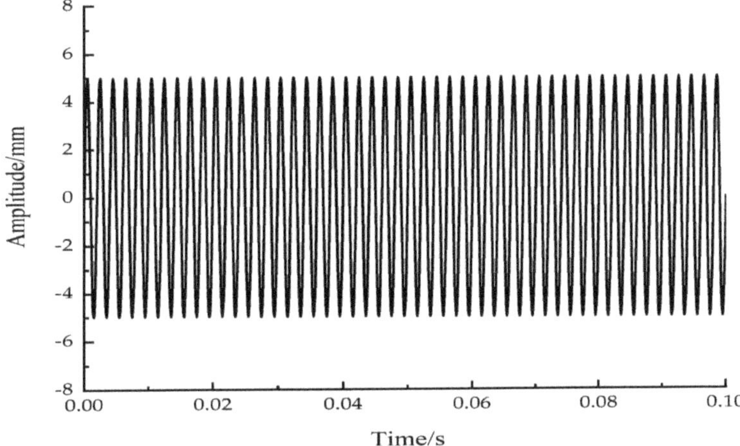

Figure 5. Waveform of the mechanical vibrations.

3. Result and Discussion

From the previous research (unpublished), the optimized technological parameters of pure PVDF (P-PVDF) fiber membrane were settled as mentioned in the Section 2.3. As shown in Figure 6a,b, the SEM images of P-PVDF fiber membranes exhibit that the continuous and fine P-PVDF fibers without any defect and beads were obtained with a certain orientation. The tensile strength, modulus of elasticity and average diameter were presented in Table 2, revealing that the prepared fibers were uniform and fine with good mechanical properties. The voltage peak exhibiting piezoelectric performance reached 270 mV in Figure 6a. It was confirmed that the β-phase of P-PVDF fibers corresponding to the peak of 20.6° (110) formed with the decrease of the peak at 18.4° (020) corresponding to α-phase of PVDF (Figure 7b), resulting in that piezoelectric property could be enhanced with the function of high voltage and stretching during the electrospinning process. All the diffraction peaks are duly indexed from the JCPDS card no. 85-1436. All the results laid the groundwork for the following enhancement research of piezoelectric performance.

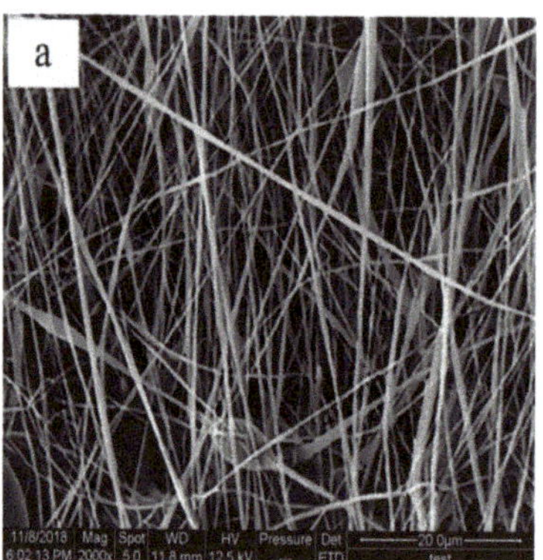
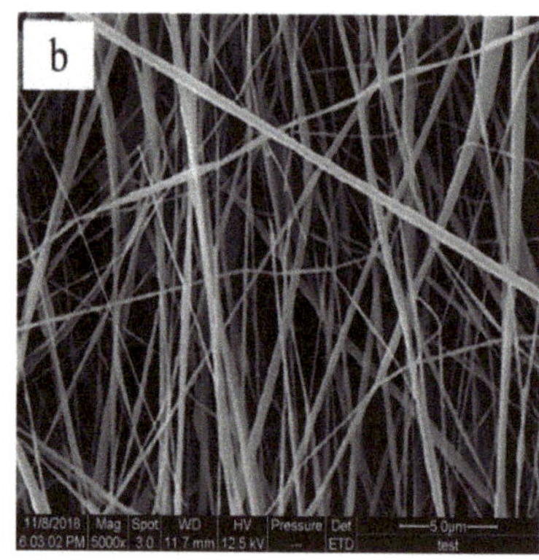

Figure 6. SEM image of the pure PVDF fiber membrane: (**a**) 2000 magnification; (**b**) 5000 magnification.

Table 2. Tensile strength, modulus of elasticity, average diameter, and diameter range of P-PVDF.

Name	Value	Standard Deviation
Tensile strength/MPa	13	0.35
Modulus of elasticity/MPa	3665	70.71
Diameter range/μm	0.1–0.7	
Average diameter/μm	0.342	
Piezoelectric signal/V	0.27	

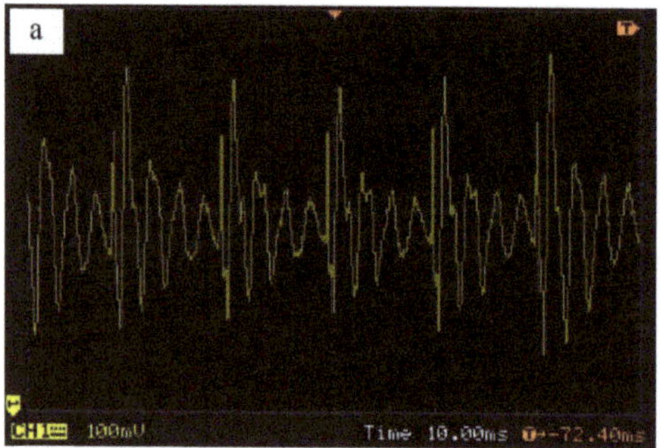
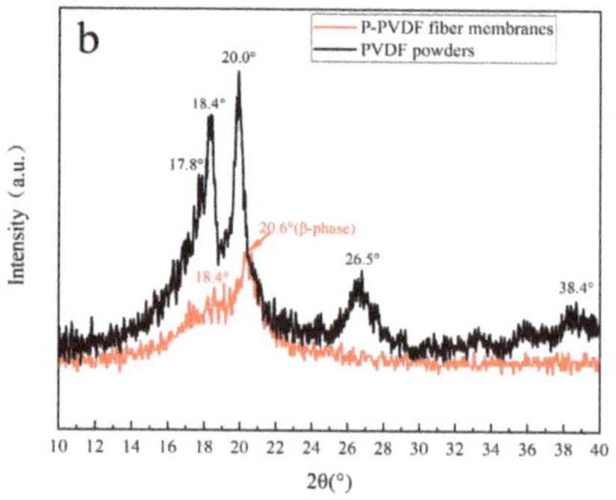

Figure 7. Piezoelectric signal image and XRD of the pure PVDF fiber membrane: (**a**) Piezoelectric signal image of the pure PVDF fiber membrane; (**b**) XRD of the pure PVDF fiber membrane.

3.1. PVDF/AgNO₃ Fiber Membrane Studies

Figure 8a–d shows the SEM images of PVDF-Ag fiber membranes at different addition of AgNO$_3$ (0.1 wt%, 0.3 wt%, 0.5 wt%, 0.7 wt%), and the corresponding fibers diameter distribution is shown in Figure 9a–d and Table 3. All the fibers are randomly arranged with a smaller average diameter (0.201 μm~0.312 μm), compared with the P-PVDF fibers (0.342 μm). With the addition of 0.3 wt%, the fine and continuous fibers exhibiting the best morphology are obtained without any beads and defects, corresponding to the minimum average diameter (0.201 μm) and the uniform fibers diameter distribution. Although the fibers with the addition of 0.1 wt% and 0.5 wt% are continuous, the uniformity of fibers decreases with the broader average diameter. When the addition is up to 0.7 wt%, the beads tare apparent with the cluttered morphology in Figure 8d, and the average diameter reach to maximum in Figure 9d. Meanwhile, along with the conductivity of PVDF/AgNO$_3$ solution, the charged droplet is directly dripped from the spinneret to result in the hard electrospinning process.

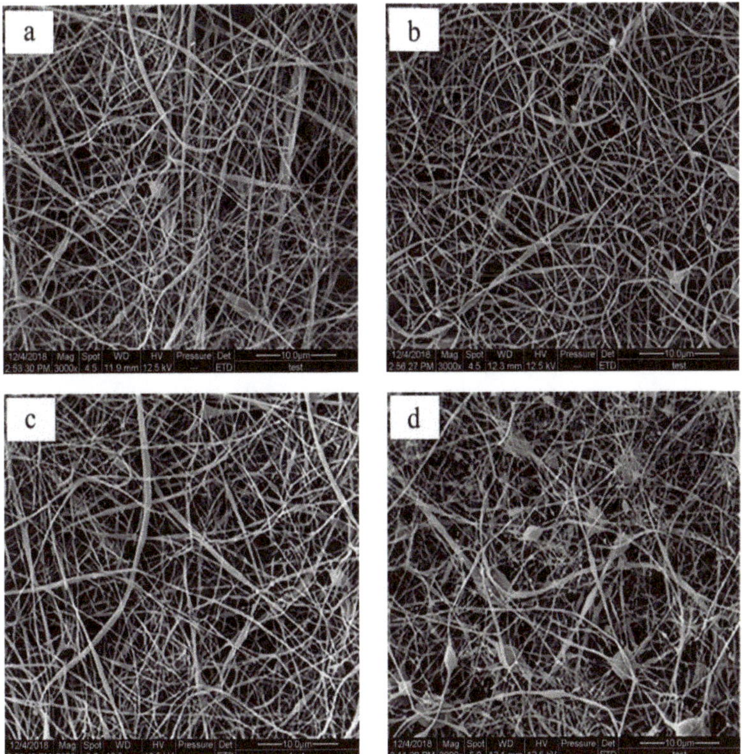

Figure 8. SEM images at different addition of AgNO$_3$: (**a**) the addition of 0.1 wt%; (**b**) the addition of 0.3 wt%; (**c**) the addition of 0.5 wt%; (**d**) the addition of 0.7 wt%.

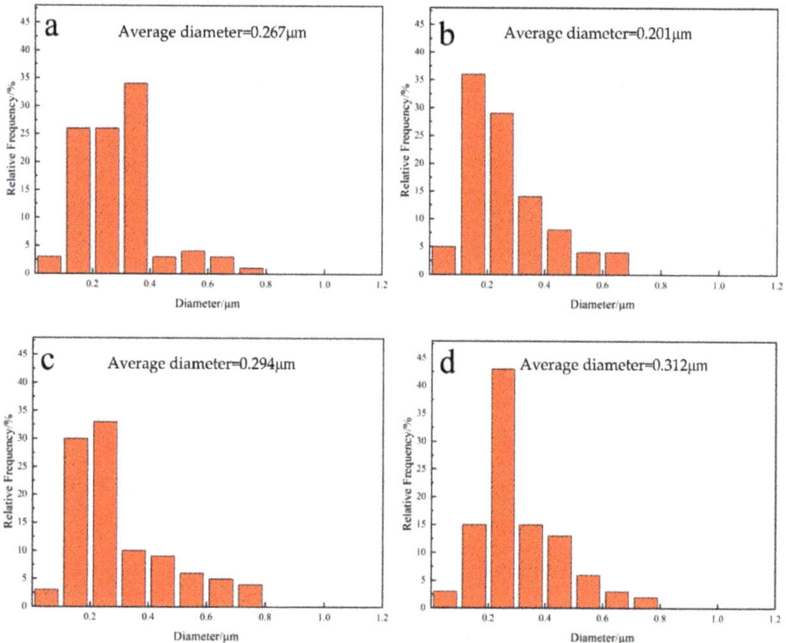

Figure 9. The fiber diameter distribution at different addition of AgNO$_3$: (**a**) the addition of 0.1 wt%; (**b**) the addition of 0.3 wt%; (**c**) the addition of 0.5 wt%; (**d**) the addition of 0.7 wt%.

Table 3. The average diameter of the fiber membrane at different loading of AgNO$_3$.

Addition/wt%	0.1	0.3	0.5	0.7	P-PVDF
Diameter distribution/μm	0–0.8	0–0.7	0–0.8	0–0.8	0.1–0.7
Average diameter/μm	0.267	0.201	0.294	0.312	0.342

The arrangement of fibers is attributed to the whipping of jet flow, which was a kind of nonaxisymmetric unstable motion affected by surface charge density. Owing to the increasing Ag$^+$ concentration of the polymer solution with the addition of AgNO$_3$, the instability of jet flow was dominated to restrain the mechanical stretching of the roller and further to generate the little steam from the jet flow. That was in favor of generating random superfine fibers to weaken the oriented characteristic of fibers. In addition, more charge accumulation during the depositional process can produce greater electrostatic repulsion, also resulting in the weak-oriented characteristic of fibers. With the addition of Ag$^+$ increasing to 0.3 wt%, the conductivity of Ag$^+$ polymer solution (Figure 1) increased, and the jet can be sufficiently stretched by the enhanced electrostatic field force to generate the finest fibers. Meanwhile, the fibers became well defined, also ascribed to the corresponding thermal conduction increased in the presence of Ag$^+$, and contributed to the improved solvent evaporation during the fly in the electrostatic field. With the addition continuing to 0.5 wt%, the rise speed of conductivity slowed down, and the speed of evaporation may be faster, and the fibers were insufficiently stretched, resulting in that the diameter of fibers increased. When the addition exceeds 0.7 wt%, the produced beads are attributed to the influence of jet flow motion, caused by the effect of the two competitions. One is that the increased surface tension of solution by the excessive inorganic salt enhances the resistance of the jet flow to make the electrospinning solution divert to liquid droplet. On the other hand, the electrostatic repulsion as the power of jet flow motion was enlarged by the increasing addition of inorganic salt to prevent the solution from clumping together into droplets. Under the two competitions, the motion of jet flow happens. When the addition increases to 0.7 wt%, the surface tension is greater than electrostatic repulsion and the drop directly sputters rather than forming jet flow. So, with the addition of $AgNO_3$ increased, the jet flow motion was enhanced and then weakened due to the greater surface tension, the same as the change of fiber morphology.

The results of piezoelectric properties are shown in Figure 10 and Table 4. Compared to the piezoelectric properties of the pure PVDF fiber membrane, the piezoelectric property of the PVDF/$AgNO_3$ fiber membrane was significantly enhanced. The voltage peak of PVDF/$AgNO_3$ fiber membranes reached 2 V with the addition of 0.3 wt%, thus attributed to the fact that the positive Ag^+ was attracted by the electronegativity $-CF_2-$ dipoles of PVDF and repelled by the $-CH_2-$ dipoles of PVDF [21]. The interaction of attraction and repulsion and enhanced interfacial polarization takes place to promote all trans confirmations and the generation of β-phase in PVDF, as shown in Figure 11. Meanwhile, with the addition of $AgNO_3$ increasing, the conductivity of polymer solution increased and corresponding the stretching in the electric field enhanced, as well as the more generated the β-phase.

However, with the addition of $AgNO_3$ increased, the voltage began to descend. The voltage was even down to 1.6 V with the addition of 0.7 wt%. It is due to that the aggregation of the Ag^+ would result in current leakage during the working process of piezoelectric signal generator. Moreover, with the addition beyond 0.3 wt%, the Ag^+ and NO_3^- were mutually and intensively attracted to each other, restricting the ions moving to the surface of jet flow in time. The inhomogeneous ions on the surface of jet flow led to weakened stretching and unstable jet flow. Therefore, the excess addition of Ag^+ and NO_3^- was not applied to the stronger piezoelectric property.

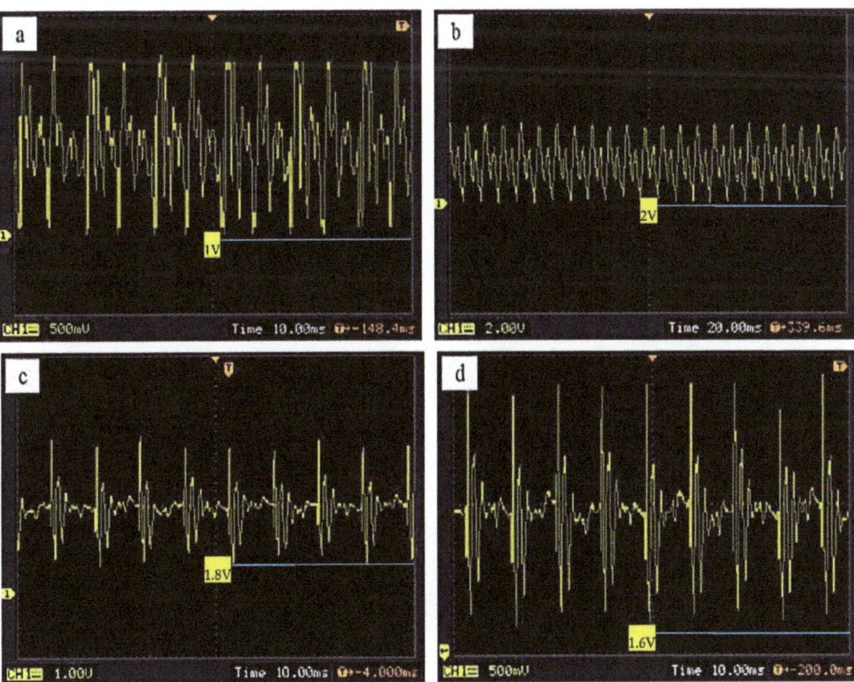

Figure 10. Piezoelectric signal of the fibers membrane at different addition of $AgNO_3$: (**a**) the addition of 0.1 wt%; (**b**) the addition of 0.3 wt%; (**c**) the addition of 0.5 wt%; (**d**) the addition of 0.7 wt%.

Table 4. Piezoelectric signal of the fiber membrane at different loading of AgNO$_3$.

Addition/wt%	0.1	0.3	0.5	0.7	**P-PVDF**
Piezoelectric signal/V	1.0	2.0	1.8	1.6	0.27

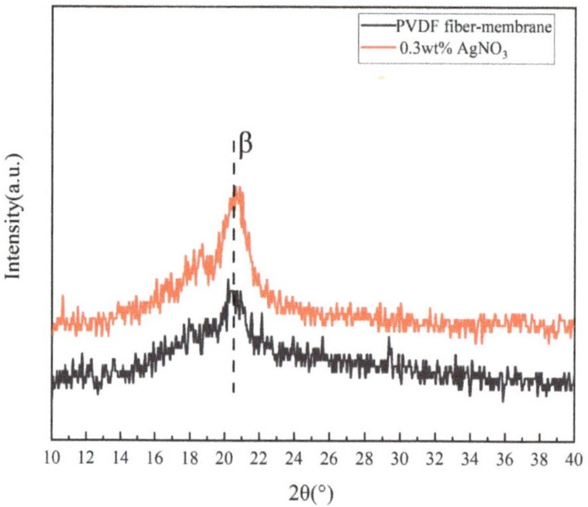

Figure 11. XRD patterns of the P-PVDF fibers membrane and PVDF/AgNO$_3$ fiber membrane at the addition of 0.3 wt% AgNO$_3$.

Table 5 shows the mechanical property of the fiber membrane at different loading of AgNO$_3$. The tensile strength and modulus of elasticity were improved by the addition of AgNO$_3$. The tensile strength and modulus of elasticity enlarged firstly and then dropped with the addition of AgNO$_3$ increased. When the load was at 0.3 wt%, the tensile strength and modulus of elasticity reached the maximum. Compared to the P-PVDF, the improved mechanical property of PVDF-Ag was contributed to the decreasing diameter of the fiber. The finer diameter affected by the stronger stretching has a higher molecular orientation, and more fiber was obtained in the same amount of spinning. Due to the fiber nearby mutually transferring energy, the tensile strength and modulus of elasticity of the fiber membrane were significantly improved.

Table 5. Mechanical properties of the fiber membrane at different loading of AgNO$_3$.

Addition/wt%	Tensile Strength/MPa	Standard Deviation	Modulus of Elasticity/Mpa	Standard Deviation
0.1	20	0.11	3652	72.89
0.3	22	0.16	4047	148.24
0.5	21	0.33	3920	96.21
0.7	20	0.19	3785	77.20
P-PVDF	13	0.35	3665	70.71

3.2. PVDF/FeCl$_3$·6H$_2$O Fiber Membrane Studies

The morphology of PVDF/FeCl$_3$·6H$_2$O fiber membrane, the average diameter and the distribution of fibers are shown in Figures 12a–d and 13a–d and Table 6, respectively. All the diameter of the fiber is fined and distributed uniformly. With the addition of 0.4 wt% and 0.8 wt%, the gap between fibers is obviously seen along with a few beads in the dense accumulation of fibers. Owing to the lower conductivity, the fibers are deposited in a certain orientation with a loading of 0.4 wt%. However, with the concentration of FeCl$_3$·6H$_2$O increased, the arrangement of fibers is absolutely random. Beyond the addition of 0.8 wt%, the diameter significantly increases from 0.254 µm to 0.399 µm. Particularly, when the addition is up to 1.6 wt%, the exhibiting adhered structure severely affects the continuity and uniformity of fibers.

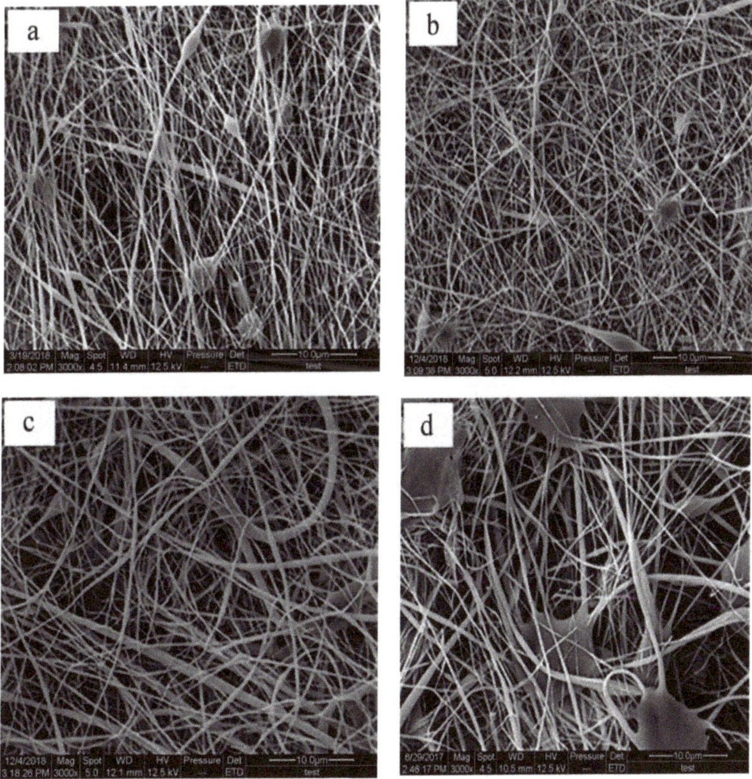

Figure 12. SEM images at different addition of FeCl$_3$·6H$_2$O: (**a**) the addition of 0.4 wt%; (**b**) the addition of 0.8 wt%; (**c**) the addition of 1.2 wt%; (**d**) the addition of 1.6 wt%.

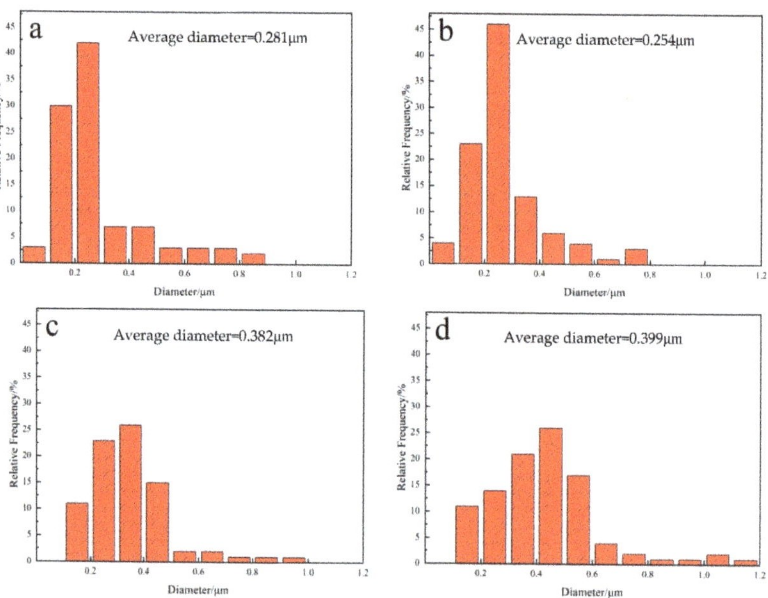

Figure 13. The fiber diameter distribution at different addition of $FeCl_3 \cdot 6H_2O$: (**a**) the addition of 0.4 wt%; (**b**) the addition of 0.8 wt%; (**c**) the addition of 1.2 wt%; (**d**) the addition of 1.6 wt%.

Table 6. The average diameter of the fiber membrane at different loading of $FeCl_3 \cdot 6H_2O$.

Addition/wt%	0.4	0.8	1.2	1.6	P-PVDF
Diameter distribution/μm	0.1–0.9	0.1–0.8	0.1–1.2	0.1–1.2	0.1–0.7
Average diameter/μm	0.281	0.254	0.382	0.399	0.342

With the addition of $FeCl_3 \cdot 6H_2O$ below the loading of 0.8 wt%, the charge in the surface of the jet flow increased with the increasing addition of $FeCl_3 \cdot 6H_2O$. Owing to that, the fine morphology of membrane representing uniform and small diameter fibers is obtained under sufficient stretching. However, when the addition continues to increase, the diameter of fibers is sharply broader. A similar phenomenon was not observed in the PVDF-Ag study. By testing the viscosity of the electrospinning solution with the different addition of $FeCl_3 \cdot 6H_2O$, the curve of viscosity is obtained in Figure 14. It was observed that the viscosity of solution decreased by the increasing addition of $FeCl_3 \cdot 6H_2O$. Especially, above the loading of 0.8 wt%, the viscosity decreasing rate of the solution increased, and the corresponding rate of conductivity slowed down. Therefore, with the flux and surface tension of the jet flow increased, the jet flow is not sufficiently stretched under the poor stretching in the electric field. In addition, the shorter duration of jet flow in the electric field with the flux increasing also broadens the fibers. Thus, the adhered structure is observed with the broadest fibers with the addition of 1.6 wt%.

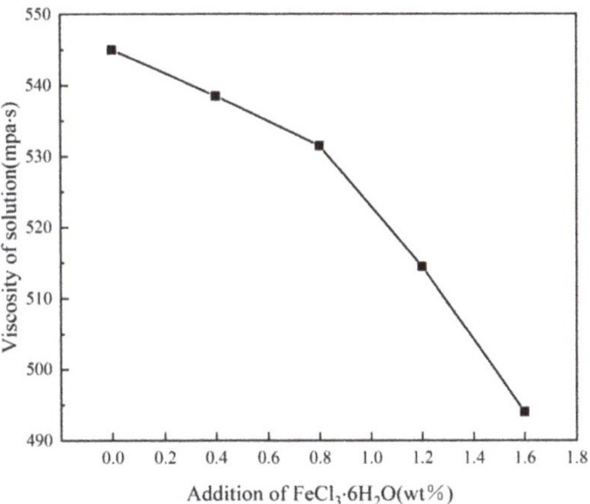

Figure 14. The viscosity of electrospinning solution at different loading of $FeCl_3 \cdot 6H_2O$.

From Figure 15 and Table 7, the piezoelectric property of PVDF-Fe fiber membrane is significantly improved. With the addition of 0.8 wt%, the peak voltage is 4.8 V and 140% higher than that of the PVDF-Ag with the addition of 0.3 wt%. However, the peak voltage obviously begins to weaken with the addition increasing. The voltage is respectively down to 2.2 V and 500 mV, corresponding to the addition of 1.2 wt% and 1.6 wt%. Meanwhile, with the excess addition of $FeCl_3 \cdot 6H_2O$, the broader diameter and the decreasing performance state that the β-phase formation is restrained by the weak stretching of jet flow.

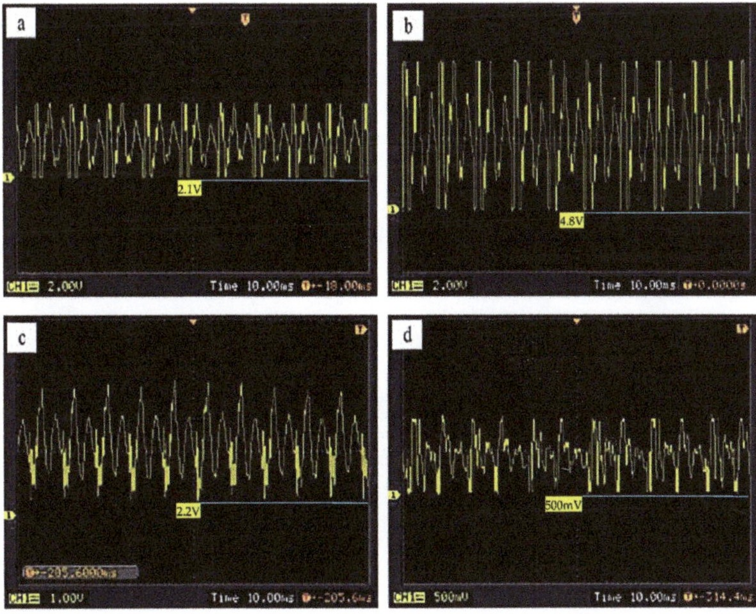

Figure 15. Piezoelectric signal of the fiber membrane at different addition of $FeCl_3 \cdot 6H_2O$: (**a**) the addition of 0.4 wt%; (**b**) the addition of 0.8 wt%; (**c**) the addition of 1.2 wt%; (**d**) the addition of 1.6 wt%.

Table 7. Piezoelectric signal of the fiber membrane at different loading of AgNO$_3$.

Addition/wt%	0.4	0.8	1.2	1.6	**P-PVDF**
Piezoelectric signal/V	2.1	4.8	2.2	0.5	0.27

As analyzed, the FeCl$_3$·6H$_2$O is a kind of covalent compound and dissolves in water part of FeCl$_3$ molecule, other than AgNO$_3$. Thus, with the same addition, the conductivit of AgNO$_3$ solution is higher than that of the FeCl$_3$·6H$_2$O solution. However, the additio of FeCl$_3$·6H$_2$O has another promoting effect except that promoting the β-phase formatio by the enhanced conductivity. The specific interactions near the Fe/PVDF interfaces ca effectively induce the nucleation of the polar (ferroelectric) phase of PVDF, promoting th polarization of PVDF and the generation of β-phase [22]. The Cl$^-$ nearby is attracted b the Fe^{3+}, which has a small ionic radius and strong polarizability, making its distributio distorted to generate electric dipole moment. Moreover, a strong electrostatic interactio between the water molecules of iron salts and the polar –CF$_2$ via the formation of hydrogen bonds may be the possible driving factor for the nucleation of polar β-phase in PVDF-Fe thin films. Owing to these, the FeCl$_3$·6H$_2$O molecules exhibit strong polarity. As known, the piezoelectric property of polar PVDF is attributed to the structure of all trans (TTTT) in the β-phase molecule generated by stretching in the electric field. With moderate FeCl$_3$·6H$_2$O added, the polarized electric dipole of FeCl$_3$·6H$_2$O by the electric field is veered and arranged regularly to promote the formation of β-phase, drastically improving the piezoelectric properties.

Figure 16 depicts the XRD patterns of P-PVDF, PVDF-Ag (0.3 wt% AgNO$_3$) and PVDF-Fe (0.8 wt% and 1.6 wt% FeCl$_3$). The peak arising at 20.6° corresponds to the β-phase of PVDF, and the highest peak corresponds to the addition of 0.8 wt% FeCl$_3$·6H$_2$O. Accordingly, the piezoelectric property is best. The reason is that the spinnability of the electrospinning solution is restricted by conductivity which depends on the addition of inorganic salt. At the same conductivity, the addition of FeCl$_3$·6H$_2$O is more than it of AgNO$_3$. Meanwhile, with the FeCl$_3$·6H$_2$O added, not only is the jet flow stretched adequately, but also the phase transformation is enhanced, exhibiting the piezoelectric properties further improved. However, with the excess addition of FeCl$_3$·6H$_2$O, the duration of jet flow in the electric field is short, resulting in the inadequate stretch and phase transformation, weakening the piezoelectric properties sharply. Therefore, the peak of PVDF-Fe (1.6 wt% FeCl$_3$·6H$_2$O) is weak, as shown in Figure 16.

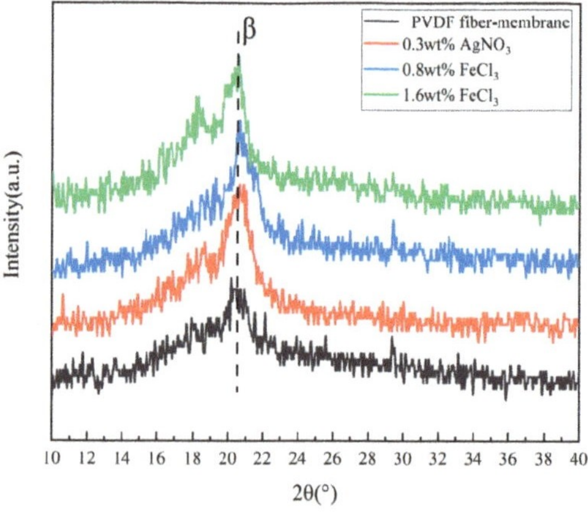

Figure 16. XRD patterns of the P-PVDF fiber membrane, PVDF/AgNO$_3$ fiber membrane at the addition of 0.3 wt%, PVDF/FeCl$_3$·6H$_2$O fiber at the addition of 0.8 wt% and 1.6 wt%.

As shown in Table 8, the mechanical properties of fiber membrane are obviously changed with the different addition of $FeCl_3 \cdot 6H_2O$. With the loading of 0.8 wt%, the tensile strength and modulus of elasticity reach the max value (19 MPa and 4056 MPa). However, with the addition increased, the tensile strength and modulus of elasticity decreased rapidly. Corresponding to the loading of 1.6 wt%, the tensile strength and modulus of elasticity are only 9 MPa and 2602 MPa, which is even lower than that of P-PVDF because of the inhomogeneous and broader fiber diameter.

Table 8. Mechanical properties of the fiber membrane at different loading of $FeCl_3 \cdot 6H_2O$.

Addition/wt%	Tensile Strength/MPa	Standard Deviation	Modulus of Elasticity/MPa	Standard Deviation
0.4	18	0.20	3301	70.99
0.8	19	0.15	4056	74.50
1.2	12	0.11	2746	63.23
1.6	9	0.09	2602	58.29
P-PVDF	13	0.35	3665	70.71

3.3. PVDF/Nanographene Fiber Membrane Studies

The nanographene has lots of excellent performance indicators, but it is easily aggregated because of the van der Waals force between carbon atoms. For the sake of making nanographene disperse more uniformly in the solution, the La rare-earth modification is adopted to modify the surface of nanographene. Figure 17 shows the dispersion of nanographene in the distilled water before and after modification. The unmodified nanographene cannot disperse well in the distilled water, but the modified one does well. The reason is that the La polarized by the non-metallic element in the rare earth modifier can clean the surface of Nanographene and form the La-C bonds to make the nanographene stable.

Figure 17. Dispersion image of nanographene in the distilled water before and after modification.

Figures 18a–d and 19a–d and Table 9 depict the SEM images of the modified PVDF-G, the average diameter, and fiber diameter distribution at different additions. It is observed that all the fibers of modified PVDF-G have uniform distribution, and the gap between fibers is obviously separated without beads. The average diameter of fibers decreases and then increases. The smoothly improved conductivity with the addition of modified nanographene contributed to the result. With the addition increasing to 1 wt%, the morphology has a great change, revealing the flattening structure and aggravated tangle of fibers. Accordingly, the average diameter reaches a minimum (0.296 μm). With the increase, the increasing conductivity accelerates the motion of graphene particles in the electric field; thus, the fibers are poorly stretched and deposited on the roller in a thicker size. Ultimately, the thicker fibers subjected to larger centrifugal force from the roller fiber form the flattening structure after the solvent evaporation. So, when the loading was up to

2 wt%, the flattening structure of fibers is obvious with the broadest diameter (0.692 μm) as shown in Figures 18d and 19d. Compared to the PVDF-Ag and PVDF-Fe, there are no beads and adhesive organizations in the PVDF-G, illustrating that the nanographene is combined with the molecular chain of PVDF. In addition, on account of conductivity, whose improvement is limited, the fiber of PVDF-G is less poorly stretched than that of PVDF-Ag and PVDF-Fe, revealing that the average diameter of PVDF-G fibers is smaller compared to the fiber of PVDF-Ag and PVDF-Fe.

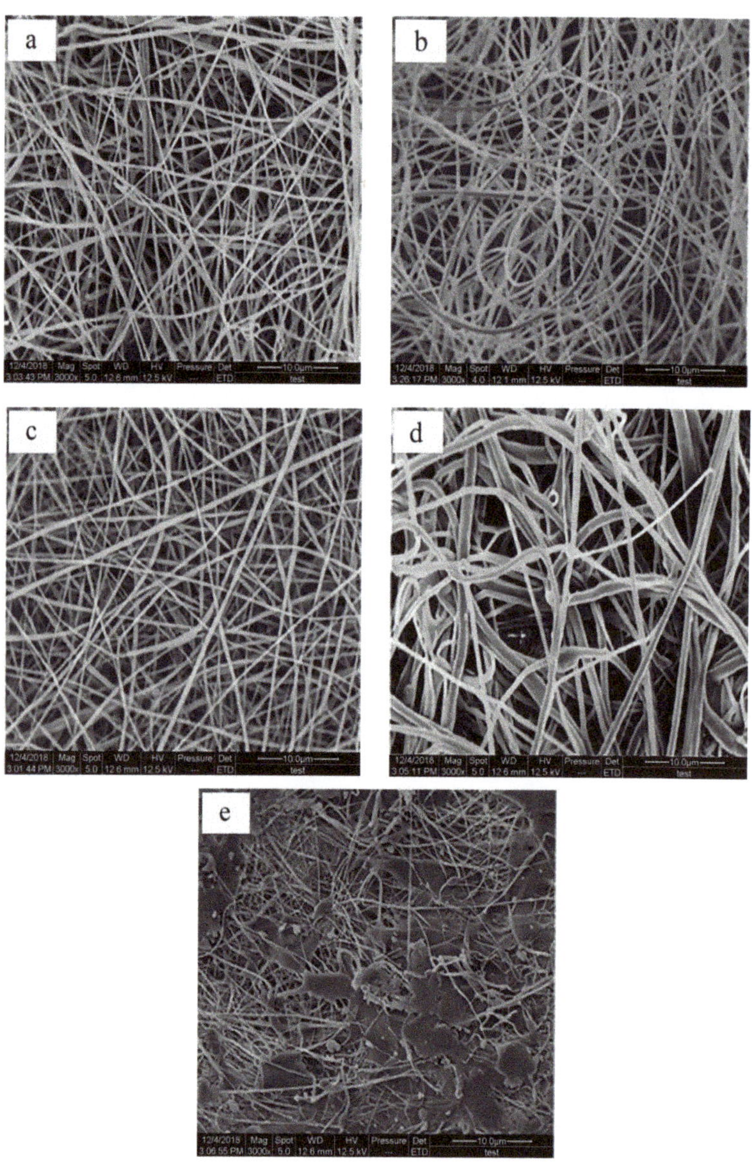

Figure 18. SEM images at different addition of nanographene: (**a**) the addition of 0.5 wt%; (**b**) the addition of 1.0 wt%; (**c**) the addition of 1.5 wt%; (**d**) the addition of 2.0 wt%; (**e**) 1 wt% unmodified nanographene.

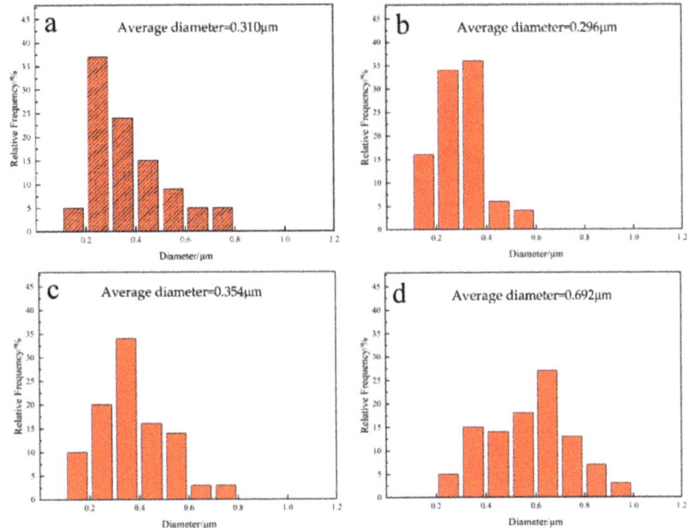

Figure 19. The fiber diameter distribution at different addition of nanographene: (**a**) the addition of 0.5 wt%; (**b**) the addition of 1.0 wt%; (**c**) the addition of 1.5 wt%; (**d**) the addition of 2.0 wt%.

Table 9. The average diameter of the fiber membrane at different loading of nanographene.

Addition/wt%	0.5	1.0	1.5	2.0	**P-PVDF**
Diameter distribution/μm	0.1–0.8	0.1–0.6	0.1–0.8	0.2–1.0	0.1–0.7
Average diameter/μm	0.310	0.296	0.354	0.692	0.342

Figure 18e shows that the fibers of unmodified PVDF-G distribute inhomogeneously without a uniform diameter of fibers. In the fibers is a mass of adhesive tissue and agglomerated unmodified nanographene absorbed on the surface of fibers. Owing to that, the fiber continuity was drastically affected to result in the large defect of fiber.

The piezoelectric signal of the fiber membrane at different loading of nanographene is shown in Figure 20a–d and Table 10. Compared to the P-PVDF, the piezoelectric property is greatly improved with the addition of modified nanographene. Corresponding to the change of diameter, the piezoelectric properties improve smoothly. With the loading increases, the peak voltage increases and then decreases. At the loading of 1.0 wt%, the peak voltage reaches the max value of 1.8 V, exhibiting the best piezoelectric property. That is illustrated by the XRD patterns of P-PVDF and PVDF-G at different loading of modified nanographene in Figure 21. It is observed that the peak of β-phase has the same variation tendency corresponding to the peak voltage. At the loading of 1.0 wt%, the strength of β-phase is strongest. The results indicate that the nanographene can promote the transformation from α-phase to β-phase. Except for promoting the stretching of jet flow by improving the conductivity of a solution, the added nanographene can also promote the formation of β-phase in other ways. One is the interface interactions. With the addition of nanographene, the partial electric field is strengthened and generates the induced charge. By the enhanced Coulombic force, the induced PVDF chains with the structure of all trans (the molecular structure of β-phase) are attracted to from crystal on the surface of nanographene. Under the interface interaction, the other phase of PVDF transforms to the β-phase. The other one is that the nanographene affects the orientation of CH_2-CF_2 electric dipole. The image (Figure 22) of the β-phase facilitated by nanographene illustrates that the CH_2-CF_2 electric dipole is oriented to F atom and closer to the nanographene particles.

By modifying nanographene, the functional group is generated on its surface. The H atoms in the functional group are tightly integrated with the F atom by the hydrogen bonding interaction. In addition, the combined action of the polar solvent (DMF) and the PVDF molecule chains affect the motion and arrangement of CH_2-CF_2 electric dipole. All these are of great benefit to the formation of β-phase and the improvement of the piezoelectric property. However, the addition of unmodified nanographene is easily aggregated and hinders the movement of PVDF molecule chains, resulting in that the peak voltage (200 mV) being lower than that of the P-PVDF fiber membrane with unmodified nanographene (Figure 20e).

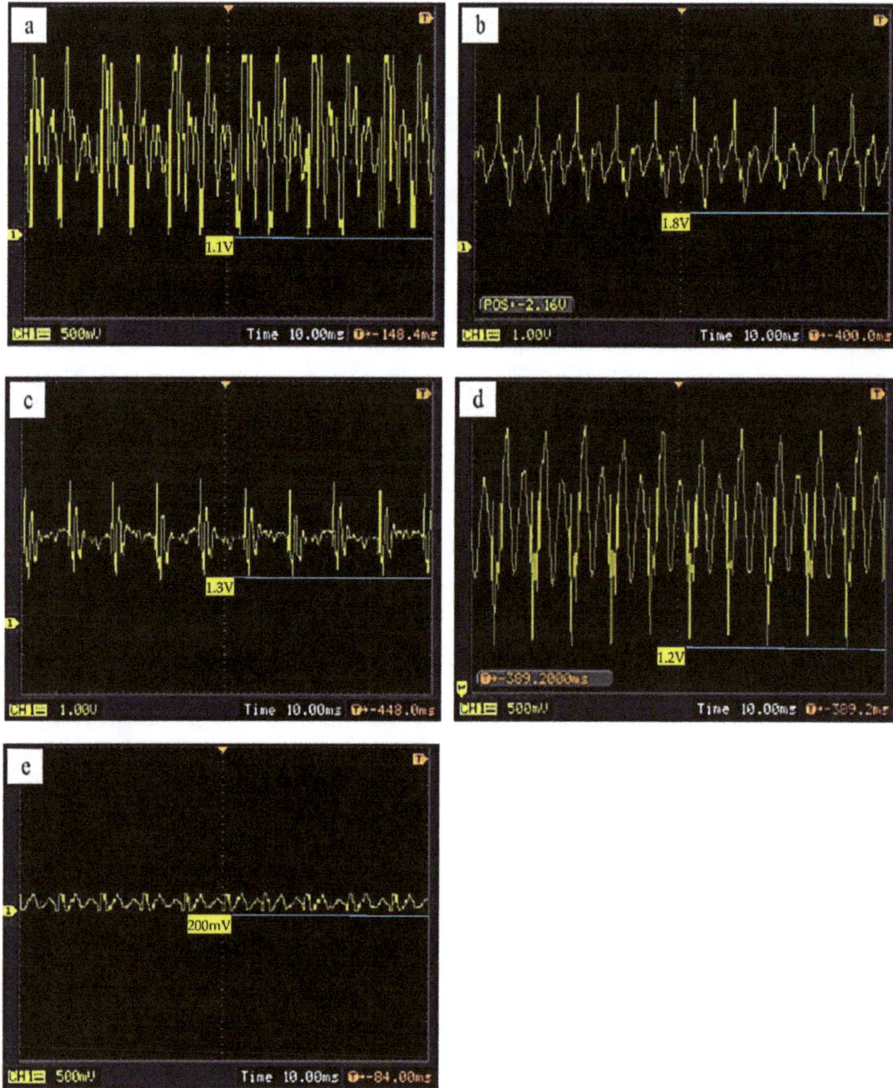

Figure 20. Piezoelectric signal of the fiber membrane at different addition of nanographene: (**a**) the addition of 0.5 wt%; (**b**) the addition of 1.0 wt%; (**c**) the addition of 1.5 wt%; (**d**) the addition of 2.0 wt%; (**e**) 1 wt% unmodified nanographene.

Table 10. Piezoelectric signal of the fiber membrane at different loading of nanographene.

Addition/wt%	0.4	0.8	1.2	1.6	1 wt% Unmodified	P-PVDF
Piezoelectric signal/V	1.1	1.8	1.3	1.2	0.2	0.27

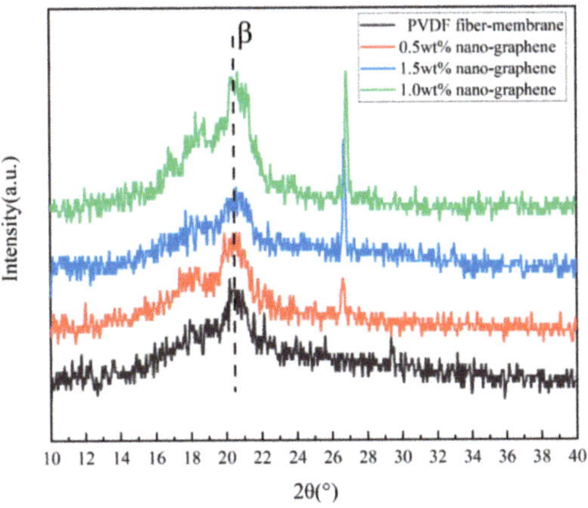

Figure 21. XRD patterns of the P-PVDF fiber membrane, PVDF/nanographene fiber membrane at the addition of 0.5 wt%, 1.5 wt%, and 2.0 wt%.

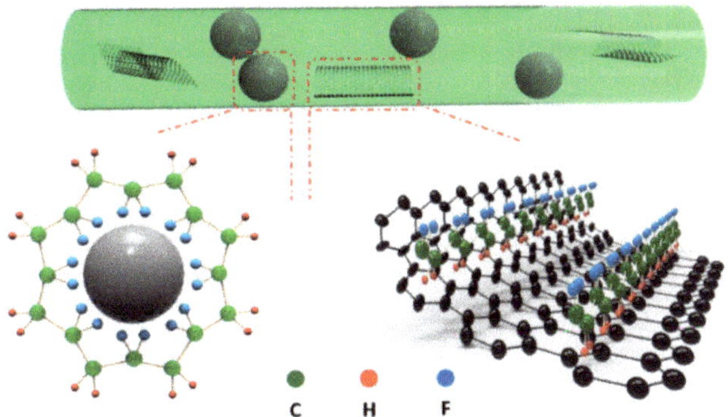

Figure 22. Image of the β-phase facilitated by nanographene.

Table 11 depicts the mechanical properties of the fiber membrane at different loading of nanographene. At the addition of 1.0 wt%, the mechanical properties are greatly improved, revealing that the tensile strength and modulus of elasticity reach to the maximum value of 31 MPa and 6538 MPa. It is illustrated that the nanographene disperses uniformly in the fibers and improves the interface combination between the PVDF and nano-reinforced phase. The tightly combined polymer chains of the nanoparticle matrix are in favor of stress transfer. Meanwhile, the C-C bonds in the nanographene can also improve the mechanical properties. However, with the addition increasing, the tensile strength and modulus of elasticity decrease. It is attributed to that the excess modified nanographene is easy to aggregate in the solution and form the defect in the membrane. And the stress concentration happened in the use process to weaken the mechanical properties. Compared with PVDF-G with modified nanographene, the tensile strength and modulus of elasticity with unmodified nanographene was only 6 MPa and 1480 MPa at the same loading, because the aggregated unmodified nanographene makes the fiber continuity weaken and form the defect on the surface of fibers.

Table 11. Mechanical properties of the fiber membrane at different loading of nanographene.

Addition/wt%	Tensile Strength/MPa	Standard Deviation	Modulus of Elasticity/MPa	Standard Deviation
0.4	22	0.15	3781	68.11
0.8	31	0.22	6583	90.19
1.2	29	0.18	5128	64.49
1.6	17	0.15	4429	73.28
Unmodified 1.0 wt%	6	0.10	1480	108.81
P-PVDF	13	0.35	3665	70.71

4. Conclusions

In summary, we successfully fabricated the composite fiber membrane with three inorganic reinforced materials ($AgNO_3$, $FeCl_3 \cdot 6H_2O$ and modified nanographene). The addition of three inorganic reinforced materials, their mechanism of action, the morphology, the mechanical and piezoelectric properties are studied. It has been found that all the addition of the inorganic reinforced materials has great benefit to morphology and the piezoelectric property improvement compared to the pure PVDF membrane. When the optimal addition of $AgNO_3$ is 0.3 wt%, the minimum average diameter, peak voltage, tensile strength and modulus of elasticity are respectively 0.201 μm, 2 V, 19.76 MPa and 4047 MPa. When the optimal addition of $FeCl_3 \cdot 6H_2O$ is 0.8 wt%, the minimum average diameter, peak voltage, tensile strength and modulus of elasticity are respectively 0.254 μm, 4.8 V, 19 MPa and 5621 MPa, where the best piezoelectric properties were obtained. When the optimal addition of modified nanographene is 1.0 wt%, the minimum average diameter, peak voltage, tensile strength and modulus of elasticity are respectively 0.296 μm, 1.8 V, 31 MPa and 6583 MPa. With the addition of modified nano-graphene, the improvement of morphology and properties is obvious, especially the mechanical properties. In addition, we demonstrate that the electroactive β-phase of PVDF induced by inorganic reinforced materials plays a crucial role in enhancing the piezoelectricity of nanocomposites. All results indicate that the addition of inorganic reinforced materials is a promising candidate for the further piezoelectric nanocomposite membrane as nucleating agents.

Author Contributions: Conceptualization, C.L.; Methodology, C.L.; Software, H.W.; Validation, C.L. and Y.F.; Formal Analysis, C.L.; Investigation, C.L., H.W., X.Y. and Q.M.; Resources, C.L.; Data Curation, Q.M. and H.C.; Writing–Original Draft Preparation, C.L. and H.W.; Writing–Review & Editing, C.L. and H.W.; Visualization, H.W.; Supervision, C.L.; Project Administration, C.L.; Funding Acquisition, C.L. All authors have read and agreed to the published version of the manuscript.

Funding: The research was funded by high technology ship research project (grand No. SSBQ-2020-HN-06-02).

Institutional Review Board Statement: Not applicable.

Informed Consent Statement: Not applicable.

Data Availability Statement: The data presented in this study are available on request from the corresponding author. The data are not publicly available as the data is a part of an ongoing study.

Acknowledgments: This work was supported by high technology ship research project (grand No. SSBQ-2020-HN-06-02).

Conflicts of Interest: The authors declare no conflict of interest.

References

1. Manish; Sukesha, S. Piezoelectric energy harvesting in wireless sensor networks. In Proceedings of the 2015 2nd International Conference on Recent Advances in Engineering & Computational Sciences (RAECS), Chandigarh, India, 21–22 December 2015; pp. 1–6.
2. Nechibvute, A.; Chawanda, A.; Luhanga, P. Piezoelectric Energy Harvesting Devices: An Alternative Energy Source for Wireless Sensors. *Smart Mater. Res.* **2012**, *2012*, 853481. [CrossRef]
3. Jung, M.; Kim, M.G.; Lee, J.-H. Micromachined ultrasonic transducer using piezoelectric PVDF film to measure the mechanical properties of bio cells. In Proceedings of the Sensors, Christchurch, New Zealand, 25–28 October 2009.
4. Shi, L.; Jin, H.; Dong, S.; Huang, S.; Kuang, H.; Xu, H.; Chen, J.; Xuan, W.; Zhang, S.; Li, S.; et al. High-performance triboelectric nanogenerator based on electrospun PVDF-graphene nanosheet composite nanofibers for energy harvesting. *Nano Energy* **2021**, *80*, 105599. [CrossRef]
5. Yang, Y.; Zhou, Y.; Zhang, H.; Liu, Y.; Lee, S.; Wang, Z.L. A Single-Electrode Based Triboelectric Nanogenerator as Self-Powered Tracking System. *Adv. Mater.* **2013**, *25*, 6594–6601. [CrossRef] [PubMed]
6. Yang, Y.; Guo, W.; Pradel, K.C.; Zhu, G.; Zhou, Y.; Zhang, Y.; Hu, Y.; Lin, L.; Wang, Z.L. Pyroelectric Nanogenerators for Harvesting Thermoelectric Energy. *Nano Lett.* **2012**, *12*, 2833–2838. [CrossRef]
7. Tang, E.; Wang, L.; Han, Y. Space debris positioning based on two-dimensional PVDF piezoelectric film sensor. *Adv. Space Res.* **2019**, *63*, 2410–2421. [CrossRef]
8. Li, Y.-J.; Wang, G.-C.; Cui, H.-Y.; Cao, S.-K.; Wang, X.-Y. Dynamic characteristics and optimization research on PVDF piezoelectric film force sensor for steel ball cold heading machine. *ISA Trans.* **2019**, *94*, 265–275. [CrossRef] [PubMed]
9. Li, Q.; Xing, J.; Shang, D.; Wang, Y. A Flow Velocity Measurement Method Based on a PVDF Piezoelectric Sensor. *Sensors* **2019**, *19*, 1657. [CrossRef] [PubMed]
10. Turdakyn, N.; Medeubaye, V.A.; Abay, I.; Adair, D.; Kalimuldina, G. Preparation of a piezoelectric PVDF sensor via electrospinning. *Mater. Today Proc.* **2021**. [CrossRef]
11. Sun, C.; Zhang, J.; Zhang, Y.; Zhao, F.; Ye, Z.G. Design and fabrication of flexible strain sensor based on ZnO-decorated PVDF via atomic layer deposition. *Appl. Surface Sci.* **2021**, *562*, 150126. [CrossRef]
12. Alluri, N.R.; Chandrasekhar, A.; Ji, H.J.; Kim, S.-J. Enhanced electroactive β-phase of the sonication-process-derived PVDF-activated carbon composite film for efficient energy conversion and a battery-free acceleration sensor. *J. Mater. Chem. C* **2017**, *5*, 4833–4844. [CrossRef]
13. Mohammadi, B.; Yousefi, A.A.; Bellah, S.M. Effect of tensile strain rate and elongation on crystalline structure and piezoelectric properties of PVDF thin films. *Polym. Test.* **2007**, *26*, 42–50. [CrossRef]
14. Hess, C.M.; Rudolph, A.R.; Reid, P.J. Imaging the Effects of Annealing on the Polymorphic Phases of Poly (vinylidene fluoride). *J. Phys. Chem. B* **2015**, *119*, 4127–4132. [CrossRef] [PubMed]
15. Lau, K.; Liu, Y.; Chen, H.; Withers, R. Effect of Annealing Temperature on the Morphology and Piezoresponse Characterisation of Poly(vinylidene fluoride-trifluoroethylene) Films via Scanning Probe Microscopy. *Adv. Condens. Matter Phys.* **2013**, *2013*, 435938. [CrossRef]
16. Ahn, Y.; Lim, J.Y.; Hong, S.M.; Lee, J.; Ha, J.; Choi, H.J.; Seo, Y. Enhanced Piezoelectric Properties of Electrospun Poly (vinylidene fluoride)/Multiwalled Carbon Nanotube Composites Due to High β-Phase Formation in Poly (vinylidene fluoride). *J. Phys. Chem. C* **2012**, *117*, 11791–11799. [CrossRef]
17. Baqeri, M.; Abolhasani, M.M.; Mozdianfard, M.R.; Guo, Q.; Oroumei, A.; Naebe, M. Influence of processing conditions on polymorphic behavior, crystallinity, and morphology of electrospun poly (VInylidene fluoride) nanofibers. *J. Appl. Polym. Sci.* **2015**, *132*. [CrossRef]
18. Yz, A.; Lw, A.; Jian, Z.A.; Qr, A.; Li, L.A.; Pc, A.; Hy, C.; Jian, G.; Xn, A. Bottom-up lithium growth guided by Ag concentration gradient in 3D PVDF framework towards stable lithium metal anode. *J. Energy Chem.* **2022**, *65*, 666–673. [CrossRef]
19. Gao, X.; Liang, S.; Ferri, A.; Huang, W.; Rou Xe, L.D.; Devaux, X.; Li, X.G.; Yang, H.; Chshiev, M.; Desfeux, R. Enhancement of ferroelectric performance in PVDF: Fe$_3$O$_4$ nanocomposite based organic multiferroic tunnel junctions. *Appl. Phys. Lett.* **2020**, *116*, 152905. [CrossRef]
20. Clausi, M.; Grasselli, S.; Malchiodi, A.; Bayer, I.S. Thermally conductive PVDF-graphene nanoplatelet (GnP) coatings. *Appl. Surf. Sci.* **2020**, *529*, 147070. [CrossRef]
21. Pusty, M.; Sinha, L.; Shirage, P.M. A flexible self-poled piezoelectric nanogenerator based on a rGO–Ag/PVDF nanocomposite. *New J. Chem.* **2019**, *43*, 284–294. [CrossRef]
22. Martins, P.; Costa, C.M.; Benelmekki, M.; Botelho, G.; Lanceros-Mendez, S. On the origin of the electroactive poly(vinylidene fluoride) β-phase nucleation by ferrite nanoparticles via surface electrostatic interactions. *CrystEngComm* **2012**, *14*, 2807–2811. [CrossRef]

Article

Investigation of Mechanical Properties for Basalt Fiber/Epoxy Resin Composites Modified with La

Chong Li [1,*], Haoyu Wang [2,*], Xiaolei Zhao [3], Yudong Fu [2], Xiaodong He [4] and Yiguo Song [1]

1   Center for Engineering Training, Harbin Engineering University, Harbin 150001, China; songyiguo@hrbeu.edu.cn
2   College of Materials Science and Chemical Engineering, Harbin Engineering University, Harbin 150001, China; fuyudong@hrbeu.edu.cn
3   China Offshore Oil Engineering Company, No. 199 Haibin 15 Road, TianjinPort Free Trade Zone, Tianjin 300461, China; zhaoxiaol@cooec.com.cn
4   National Key Laboratory of Science and Technology on Advanced Composites in Special Environment, Harbin Institute of Technology, Harbin 150001, China; hexd@hit.edu.cn
*   Correspondence: lichong@hrbeu.edu.cn (C.L.); haoyuwang@hrbeu.edu.cn (H.W.)

Abstract: As an efficient reinforcing material in resin matric composites, the application of basalt fibers (BFs) in composites is limited by the poor interfacial adhesion between BFs and the resin matrix. In this study, to obtain the basalt fibers/epoxy resin composites with enhanced mechanical properties, the modification solution containing different concentrations of Lanthanum ions (La^{3+}) was synthesized to modify the BFs surfaces to enhance the poor interfacial adhesion between BFs and the matrix. The morphology, the chemical structure and the chemical composition of the modified BFs surface were observed and detected by scanning electron microscopy, Fourier transform infrared spectroscopy and X-ray photoelectron spectroscopy, respectively. The results show that, after BFs were soaked in the modification solution, the more active groups (C=O, –OH, C–O, etc.) were introduced to the BFs surfaces and effectively enhanced the bond strength between BFs and the resin matrix. The obtained mechanical performances of prepared basalt fibers/epoxy resin composites showed that the tensile strength, bending strength and interlaminar shear strength (ILSS) were improved with the modified BFs, and reached to 458.7, 556.7 and 16.77 Mpa with the 0.5 wt.% La. Finally, the enhancement mechanism of the modification solution containing La element is analyzed.

Keywords: basalt fibers; composites; mechanical properties; surface engineering

Citation: Li, C.; Wang, H.; Zhao, X.; Fu, Y.; He, X.; Song, Y. Investigation of Mechanical Properties for Basalt Fiber/Epoxy Resin Composites Modified with La. *Coatings* **2021**, *11*, 666. https://doi.org/10.3390/coatings11060666

Academic Editor: Maude Jimenez

Received: 22 April 2021
Accepted: 29 May 2021
Published: 31 May 2021

Publisher's Note: MDPI stays neutral with regard to jurisdictional claims in published maps and institutional affiliations.

Copyright: © 2021 by the authors. Licensee MDPI, Basel, Switzerland. This article is an open access article distributed under the terms and conditions of the Creative Commons Attribution (CC BY) license (https:// creativecommons.org/licenses/by/ 4.0/).

1. Introduction

In the past years, many studies have been conducted on the applicability of basalt fibers (BFs). As a reinforcing material for composites, BFs (mainly composed by SiO_2) can be prepared from basalt rocks using conventional equipment with the advantage on the low-cost process of melting and drawing [1]. Compared with other fiber materials, BFs have the excellent properties such as high tensile strength, high E-modulus, high abrasion strength, high temperature resistance, high resistance to aggressive media, excellent thermal and sound insulation, good chemical stability [2]. Moreover, the mechanical properties of BFs can be kept without significant decrease in the operating temperature range of 200–600 °C [3]. In addition, the environmentally friendly preparation and recycling process of BFs significantly improves its demand for the modern market [1,4–6]. Thus, the BFs might be a promising reinforced material as the candidate of glass and carbon fibers in composite materials, and have been attractive in the application of friction materials, corrosion resistance materials, heat shields, thermal insulting barriers and hot fluid transportation pipes in many fields [7].

Especially in the field of flywheel rotors' structural materials, the fiber-reinforced resin matrix composite flywheel was generally to replace the traditional metal flywheel due to their high strength, high energy storage density and low specific volume [8]. Compared with the common reinforced fibers (glass and carbon fibers), the BFs represent an alternative as the substitution due to their excellent properties.

However, the smooth and chemically inert surface of BFs [9–11] results in the low interfacial adhesion between BFs and matrix (here below referred as interfacial adhesion) to reduce the composite materials' performance. To enhance the interfacial adhesion, many studies have been conducted to modify the fiber surface, such as the plasma modification technology, the oxidized modification technology, the coating modification technology, and so on [12–23]. Manikandan and Jain et al. [3,24] increased the surface roughness of BFs by the acid and alkali chemical treatments, which improved interfacial bonding strength between fibers and matrix. Kim et al. [11] and Ricciardi [25] used surface treatment of BFs by low-temperature atmospheric oxygen plasma and SF_6 plasma to increase the wettability of BFs remarkably, and improve adhesive force between fiber/resin interfaces accompanied by physical etching and by the formation of chemical functional groups containing oxygen and nitrogen on the fiber surface. This increased the interlaminar fracture toughness of basalt/epoxy woven composites. Wei et al. proposed the coating modification as an effective way in improving the mechanical properties of BFs and the properties of basalt fiber/epoxy resin composites [1,6]. Cheng et al. [26–28] investigated the effects of surface-treated F-12 aramid fibers and carbon fibers on the interfacial adhesion between fibers and matrix. They found that rare earth modification solution treatment can increase the concentration of reactive functional groups on fiber surface through chemical coordinating reaction. Additionally, the interfacial adhesion was promoted obviously and the tensile properties of composites improved significantly. Meanwhile, the tensile strengths of single fibers are almost not affected by rare earth modification solution treatment. It exhibits an attractive method to treat BFs with the rare earth modification solution due to its high efficiency, simple process, environmentally friendly characteristics, and preservation of fiber tensile strengths. Thus, with the few studies on BFs modified by rare earth elements, the modification study on BFs to prepare fiber-reinforced composites with excellent performance was promising.

In this study, the rare earth modification solution was synthesized by adding different concentrations of element La. The effects of the BFs surfaces with different concentration of element La were investigated. The mechanical properties of the composite materials, such as the tensile strength, the bending strength and the interlaminar shear strength (ILSS) were detected. The modification mechanism of the BFs surface was also discussed.

2. Experimental Details

2.1. Materials

The used BFs cloth (223 g/km, unidirectional cloth) with an average fiber of diameter (5.9 μm) was manufactured by Shanxi Basalt Fiber Technology Co., Ltd., Taiyuan, China. The elastic modulus and the tensile strength of the BFs cloth were about 7.5 and 1835 MPa, respectively. The epoxy resin (LY564) and the curing agent (22964, aliphatic polyamine curing agent) were from Huntsman Co., Ltd. (The Woodlands, TX, USA). The following chemical reagents were used: acetone and nitric acid (analytical reagent, provided by Tianjin Tianda Chemical Reagent Factory, Tianjin, China), absolute ethyl alcohol (analytical reagent, produced by Tianjin Tianli Chemical Reagent Co., Ltd., Tianjin, China), citric acid and urea (analytical reagent, provided by Tianjin Zhiyuan Chemical Reagent Co., Ltd., Tianjin, China), lanthanum chloride (99.99%, manufactured by Jining Zhongkai New Type Material Co., Ltd., Jining, China).

2.2. Treatment of Fibers

The BFs cloth was cut into 260 × 35 mm². After immersed in acetone for 48 h at room temperature to remove the protective coating on the surface of BFs, the fibers were dried in a drying oven at 100 °C for 1 h. The dried fiber cloth was soaked in the concentrated nitric acid with constant temperature water bath at 60 °C for 2 h to improve the roughness and the –OH numbers of the fibers' surfaces [29]. Washing the fibers 3–4 times with distilled water to remove acid fluid from their surfaces, the PH value on the surface of the fibers was adjusted to 7. Finally, the fibers were dried in a drying oven at 100 °C for 1 h and conserved in the dryer.

According to a certain weight ratio, the composition of the BFs modification solution were uniformly mixed at room temperature, as shown in Table 1. Additionally, five kinds of rare earth modification solution were prepared with different La^{3+} concentrations of 0.1, 0.3, 0.5, 0.7, and 0.9 wt.% After immersed in the rare earth modification solution for 2 h, the treated BFs were dried in the drying oven at 100 °C for 1 h.

Table 1. Composition of the BFs modification solution.

Name	Molecular Formula
Ethyl alcohol	C_2H_6O
Citric acid	$C_6H_8O_7$
Urea	$CO(NH_2)_2$
Nitric acid	HNO_3
Lanthanum chloride	$LaCl_3 \cdot 7H_2O$

2.3. Fabrication and Measurements of Basalt Fiber/Epoxy Resin Composites

The prepared BFs was immersed in epoxy resin/curing agent mixture at volume ratio 4:1. With the hand lay-up method, the BF/ERCs were paved as three layers in the mold, and fabricated after curing in a drying oven at 120 °C for 15 min and 140 °C for 2 h, respectively.

The change of BFs' chemical structure was determined in the transmission mode using Fourier transform infrared spectroscopy (FT-IR, Perkin Elmer 100, Waltham, MA, USA). X-ray photoelectron spectroscopy (XPS) using a PHI 5700 ESCA System equipped with an Al-Ka X-ray source (NewYork, NY, USA), analyzed the chemical composition of the BFs surfaces. The surface morphologies of BFs and the fracture surface morphologies of the composites were examined using scanning electron microscopy (SEM, JSM-6480, Tokyo, Japan).

The tensile test and the bending test were carried out to determine the tensile strength and the bending strength of BF/ERCs, respectively. The ILSS of BF/ERCs was measured by the three-point short-beam bending test. Due to the details of the mechanical tests shown in Table 2 below, the specimens were cut according to the following dimensional requirement.

Table 2. Details of the mechanical tests.

Test	Standard Test Method	Equipment	Loading Speed mm/min	Thickness (mm)	Width (mm)	Span (mm)
Tensile	GB/T1447-2005	Universal testing machine (UTM, WOW-50, Jinan Liangong Testing Technology Co., Ltd., Jinan, China)	5.0	1.5–3	25	100 ± 0.5
Flexural	GB/T1449-2005		constant	1.5–3	15 ± 0.5	22.5
Shear	GB3357-82		2.0	2.5	10	12.5

3. Results

3.1. FT-IR Spectra of BFs

The chemical structures of untreated and treated BFs were obtained with FT-IR spectroscopy, as shown in Figure 1. The characteristic peaks of –OH at 3450 cm^{-1} and C=O at 1630 cm^{-1} were observed in the untreated BFs. In addition, the peaks of –CH$_2$ appeared at 2921 and 2844 cm^{-1}, respectively. As the main composition of BFs, the peak of Si-O from SiO$_2$ of BFs appeared at 860 cm^{-1}.

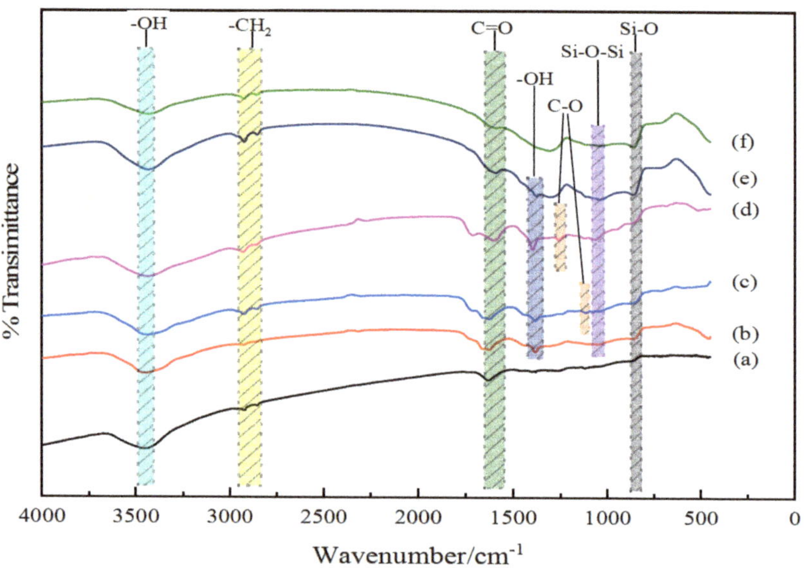

Figure 1. FT-IR spectra of BFs: (**a**) untreated; (**b**) treated with modification solution containing La^{3+} 0.1 wt.%; (**c**) treated with modification solution containing La^{3+} 0.3 wt.%; (**d**) treated with modification solution containing La^{3+} 0.5 wt.%; (**e**) treated with modification solution containing La^{3+} 0.7 wt.%; (**f**) treated with modification solution containing La^{3+} 0.9 wt.%.

For BFs treated with the 0.1 wt.% La^{3+} rare earth modification solution, the new absorption peaks of C=O at 1660 cm^{-1} and –OH at 1384 cm^{-1} appeared in the range of 4000–1330 cm^{-1}, and the peaks of –CH$_2$ at 2917 and 2845 cm^{-1} from the citric acid were broader than those of the untreated BFs, indicating that the La^{3+} was successfully absorbed on the BFs' surface with the introduced –CH$_2$– from the acyl group and citric acid, which also brought C=O. The absorption band at the peak of 1046 cm^{-1} referred to the formation of along with the peak of shifting from 860 to 863cm^{-1}, stating that the Si–OH from the hydrolysis of SiO$_2$ react with itself or the introduced –COOH under dehydration condensation reaction to produce Si–O–Si and more Si–O. With La^{3+} concentration increasing to 0.3 wt.% in the rare earth modification solution, the new peak of C–O from the citric acid along with the introduced La^{3+} at 1116 cm^{-1} appeared in the FT-IR spectrum. Moreover, the peaks of –CH$_2$ at 2922 and 2859 cm^{-1} exhibited obviously. It indicated that the La^{3+} was absorbed to the surface with –COOH and –CH$_2$ from citric acid. When La^{3+} concentration was up to 0.5 wt.%, the vibration frequency of C–O increased from 1116 to 1260 cm^{-1}, and that of C=O stretched from 1668 to 1716 cm^{-1}, accompanied by the urea in the modification solution. Due to the strong affinity of La^{3+} to nonmetal element, the La^{3+} could combine with urea and citric acid under the coordination reaction to introduce the C=O and C–O from the modification solution, resulting that more active oxygen-contained functional groups were introduced onto the fiber surfaces with the increasing La^{3+} concentration.

With the La^{3+} concentration above 0.5 wt.%, the peak of C=O began to disappear. At the La^{3+} concentration of 0.9 wt.%, the peak disappearances of C–O at 1260 cm^{-1} and the other weakened characteristic peaks were observed, implying that superabundant active groups introduced by the excess La^{3+} were aggregated on the BFs to be supersaturated and further destroyed formed macromolecular membrane on the BFs' surface to trigger the loss of active groups.

3.2. XPS Results of BFs

3.2.1. XPS Results of BFs

As shown in Figure 2 and Table 3, the XPS survey spectra and the detected atomic concentration of the elements C, O, N and Si on the BFs surfaces were respectively obtained to analyze the chemical composition of the BFs surface. On the surfaces of BFs treated with the rare earth modification solution, the concentration of element C decreased and then increased with the increase of La^{3+} concentration, but the changes of the elements O and N were contrary. The proportions of O/C and N/C on the BFs surfaces had the same variation trend with the concentration of element C. When La^{3+} concentration was 0.5 wt.%, the proportions of O/C and N/C reached maximum values, confirming that the BFs surfaces contained the largest number of active oxygen-containing groups.

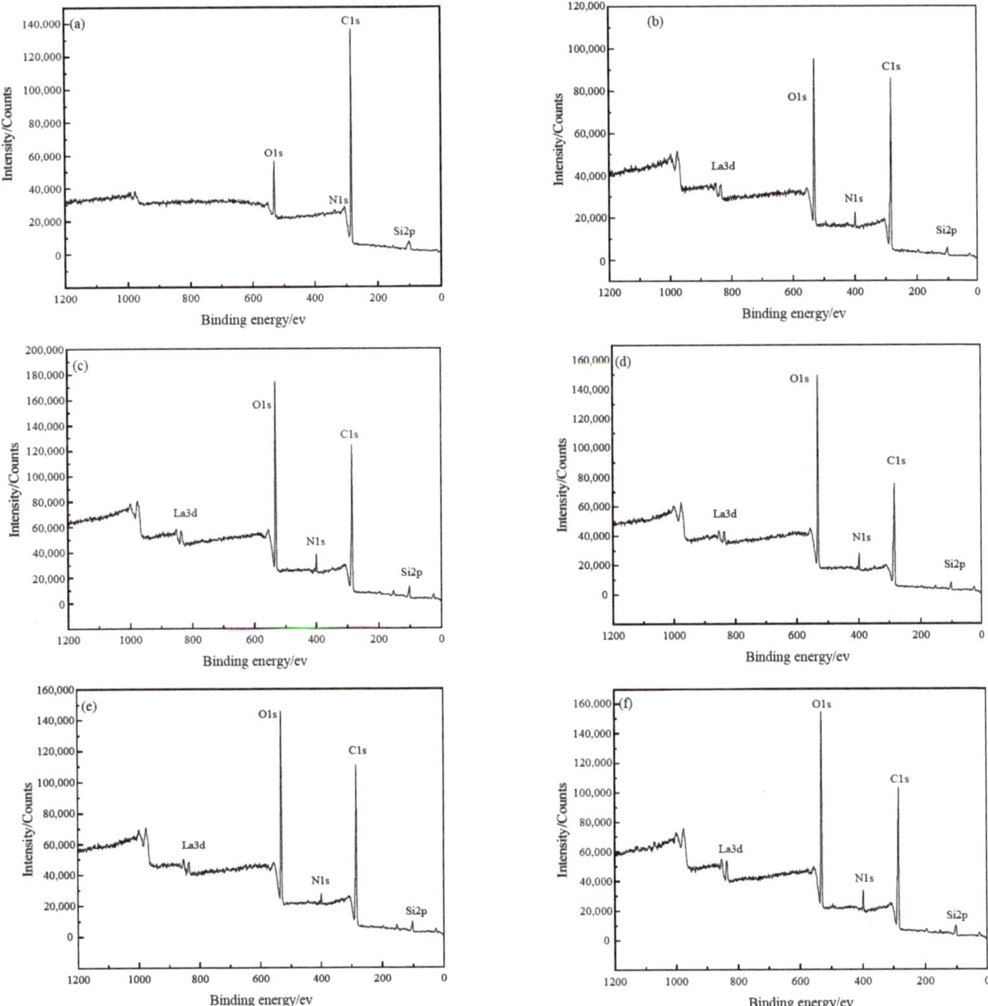

Figure 2. XPS survey spectra of BFs: (**a**) untreated; (**b**) treated with modification solution containing La^{3+} 0.1 wt.%; (**c**) treated with modification solution containing La^{3+} 0.3 wt.%; (**d**) treated with modification solution containing La^{3+} 0.5 wt.%; (**e**) treated with modification solution containing La^{3+} 0.7 wt.%; (**f**) treated with modification solution containing La^{3+} 0.9 wt.%.

Table 3. XPS results of BFs.

Specimens	Elements (%)					O/C	N/C
	C	O	Si	N	La		
Untreated	80.65	15.06	2.47	0.78	-	0.19	0.01
0.1 wt.% La^{3+}	65.29	25.55	5.38	3.72	0.07	0.39	0.06
0.3 wt.% La^{3+}	56.14	28.62	5.59	7.63	0.22	0.51	0.14
0.5 wt.% La^{3+}	40.95	43.38	5.78	9.01	0.88	1.05	0.22
0.7 wt.% La^{3+}	50.42	33.97	6.31	8.36	0.94	0.67	0.17
0.9 wt.% La^{3+}	53.38	28.68	8.15	8.32	1.47	0.54	0.16

3.2.2. XPS Results of La on BFs Surfaces

The fitted curves of La3d spectra on BFs surface treated with different modification solution containing La^{3+} are presented in Figure 3. The spectra peaks of La3d in the treated BFs with the different La^{3+} modification solution concentrations (0.1, 0.3, 0.5, 0.7, and 0.9 wt.%) were corresponding to the bonding energies of 851.02, 851.77, 851.9, 852, and 852.1 ev, respectively. All the bonding energies were lower than that (853.0 ev) of the La corresponding to La3d in $LaCl_3$, indicating that the coordination chemical reactions occurred between the La and the O, C, N element of the treated fibers' surfaces to the generate lanthanum coordination compounds on fibers' surfaces.

3.2.3. XPS Characterization of Element C on BFs Surfaces

The fitted curves of C 1s spectra and the detected functional groups contents on the BFs surfaces are presented in Figure 4 and Table 4, respectively. Three small peaks of C–C, C–O– and C=O appeared under the C 1s peak of the untreated BFs surfaces in Figure 4a, indicating that C atoms were combined with O atoms. The percentage of C atoms in the corresponding chemical state was represented with the area ratio of every small peak to the C 1s peak. Due to the low content of element N on BFs surfaces, the C–N bond could be ignored.

As seen in Figure 4b, with the peak appearance of –COOH in the fitted curves of C 1s spectra of the BFs, new functional groups were generated on the surface of BFs treated with the rare earth modification solution, besides the original ones. According to the low-to-high order of binding energy, these functional groups can be arranged as alcoholic hydroxyl (C–OH) or ether bond (C–O–C), carbonyl (C=O) and carboxyl (COOH). The –C–C content on the BFs surfaces firstly decreased and then increased with La^{3+} concentration increasing, but the changes of the carbon–oxygen bond were contrary. When La^{3+} concentration was 0.5 wt.%, with the sum of C–OH/–C–O–C–, C=O and –COOH reaching to a maximum value (72.87%), the number of oxygen-containing groups on the BFs surfaces were the largest. As La^{3+} concentration exceeded 0.5 wt.%, the content of carbon–oxygen bond decreased. That was attributed to the fact that the abundant oxygen-containing groups introduced by excess La were aggregated on the fibers' surfaces and could not be stably absorbed to the fibers' surfaces with the weak Van der Waals force between La^{3+}. Moreover, the aggregated supersaturated active oxygen-containing groups could also trigger the break of the origin formed macromolecular membrane on the fibers' surfaces, which led to the loss of active groups.

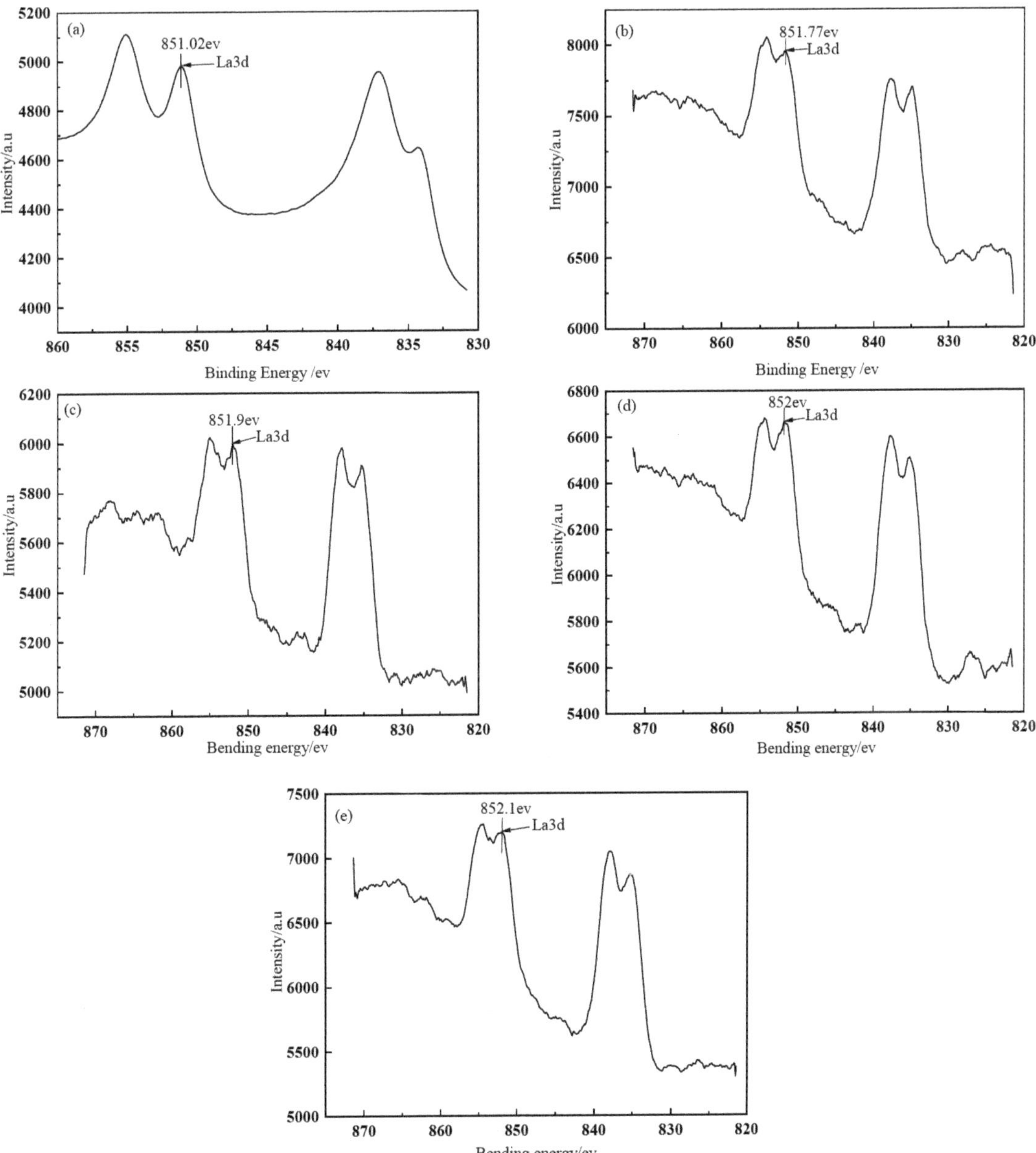

Figure 3. XPS survey spectra of La on BFs surfaces: (**a**) treated with modification solution containing La^{3+} 0.1 wt.%; (**b**) treated with modification solution containing La^{3+} 0.3 wt.%; (**c**) treated with modification solution containing La^{3+} 0.5 wt.%; (**d**) treated with modification solution containing La^{3+} 0.7 wt.%; (**e**) treated with modification solution containing La^{3+} 0.9 wt.%.

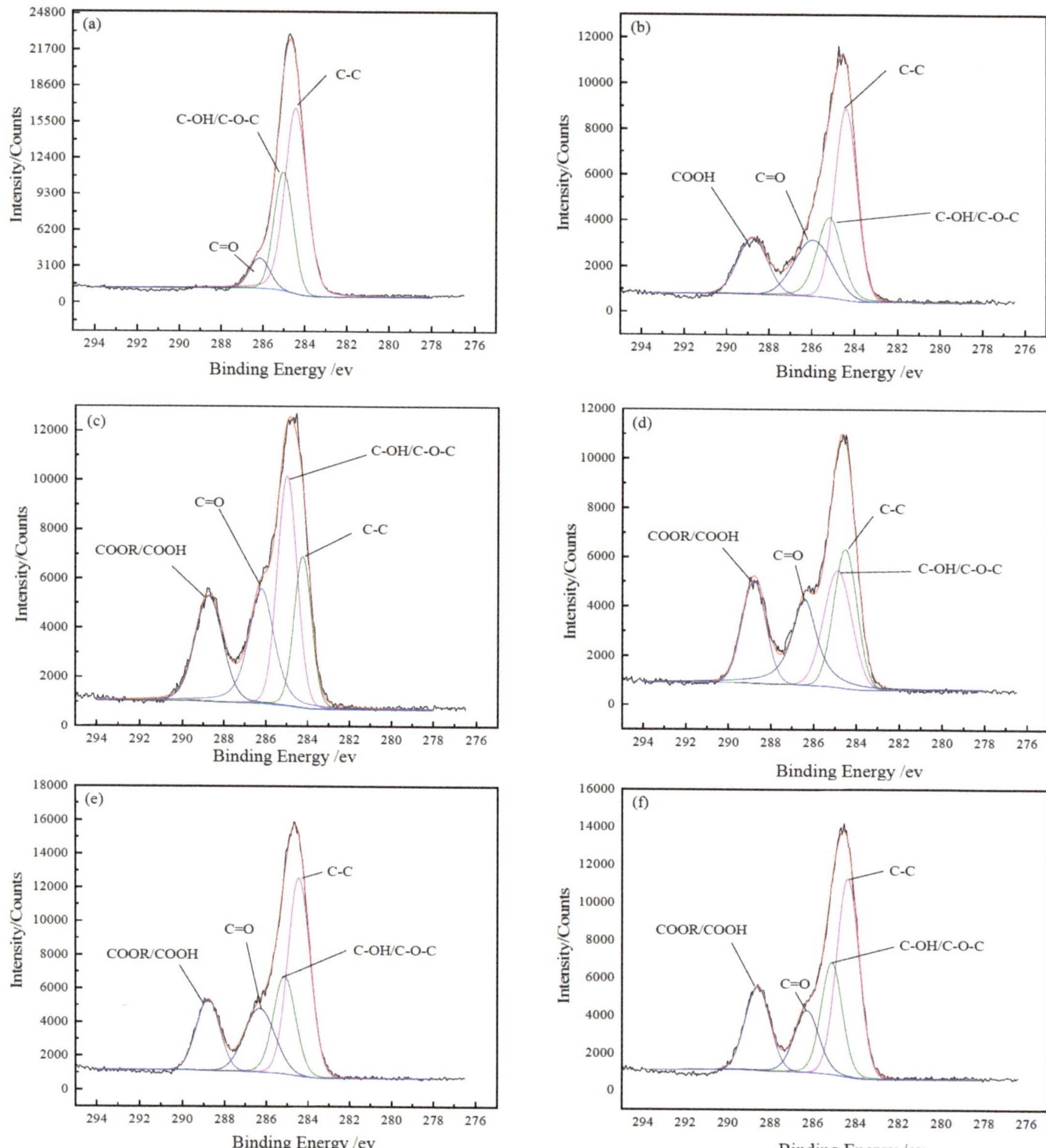

Figure 4. Fitted curves of C1s spectra: (**a**) untreated; (**b**) treated with modification solution containing La^{3+} 0.1 wt.%; (**c**) treated with modification solution containing La^{3+} 0.3 wt.%; (**d**) treated with modification solution containing La^{3+} 0.5 wt.%; (**e**) treated with modification solution containing La^{3+} 0.7 wt.%; (**f**) treated with modification solution containing La^{3+} 0.9 wt.%.

Table 4. C 1s results of BFs surfaces.

BFs	–C–C		C–OH/–C–O–C–		C=O		–COOH	
	Area Ratio (%)	Binding Energy (ev)	Area Ratio (%)	Binding Energy (ev)	Area Ratio (%)	Binding Energy (ev)	Area Ratio (%)	Binding Energy (ev)
Untreated	60.49	284.41	31.12	285	8.40	286.15	-	-
0.1 wt.% La^{3+}	41.89	284.33	21.23	285.15	20.56	286.473	16.32	288.89
0.3 wt.% La^{3+}	30.41	284.23	30.98	284.88	23.37	286.1	15.23	288.66
0.5 wt.% La^{3+}	27.13	284.52	27.36	284.87	24.56	286.43	20.96	288.8
0.7 wt.% La^{3+}	39.23	284.45	22.76	285.3	17.80	286.44	20.21	288.7
0.9 wt.% La^{3+}	42.18	284.37	22.98	285.11	15.35	286.29	19.49	288.6

3.2.4. XPS Characterization of Si Element on BFs Surfaces

The functional group contents on the BFs surfaces are presented in Table 5, according to the fitted curves of Si 2p spectra in Figure 5. Figure 5a showed that SiO_2 and Si–O–Si bonds appeared on the untreated BFs surfaces, affirming that SiO_2 was the main chemical composition of BFs surfaces. The existence of Si–O–Si bonds was ascribed to the formation of BFs connected by $[SiO_4]$ tetrahedron, which could further form $[Si_2O_7]$ under interaction and even the chain structure with a higher degree of polymerization. Under the hydrolysis of SiO_2 in the modification solution, the new functional group (Si–OH) was generated on the surface of BFs treated with the rare earth modification solution along with the obvious decrease of SiO_2. However, the increase of Si–O–Si was mainly ascribed to the dehydration condensation reaction of Si–OH bonds by themselves or active organic groups in the modification solution, such as –COOH and –OH.

Table 5. Si 2p results of BFs surfaces.

BFs	SiO_2		Si–O–Si		Si–OH	
	Area Ratio (%)	Binding Energy (ev)	Area Ratio (%)	Binding Energy (ev)	Area Ratio (%)	Binding Energy (ev)
Untreated	71.14	102.59	28.86	101.75	-	-
0.1 wt.% La^{3+}	47.41	106.16	33.64	103.48	18.94	102.43
0.3 wt.% La^{3+}	40.15	105.85	36.81	103.37	23.04	102.47
0.5 wt.% La^{3+}	36.12	105.81	39.72	103.28	24.17	102.33
0.7 wt.% La^{3+}	51.80	105.97	27.32	103.25	20.88	102.18
0.9 wt.% La^{3+}	58.13	105.71	23.89	103.17	17.98	102.28

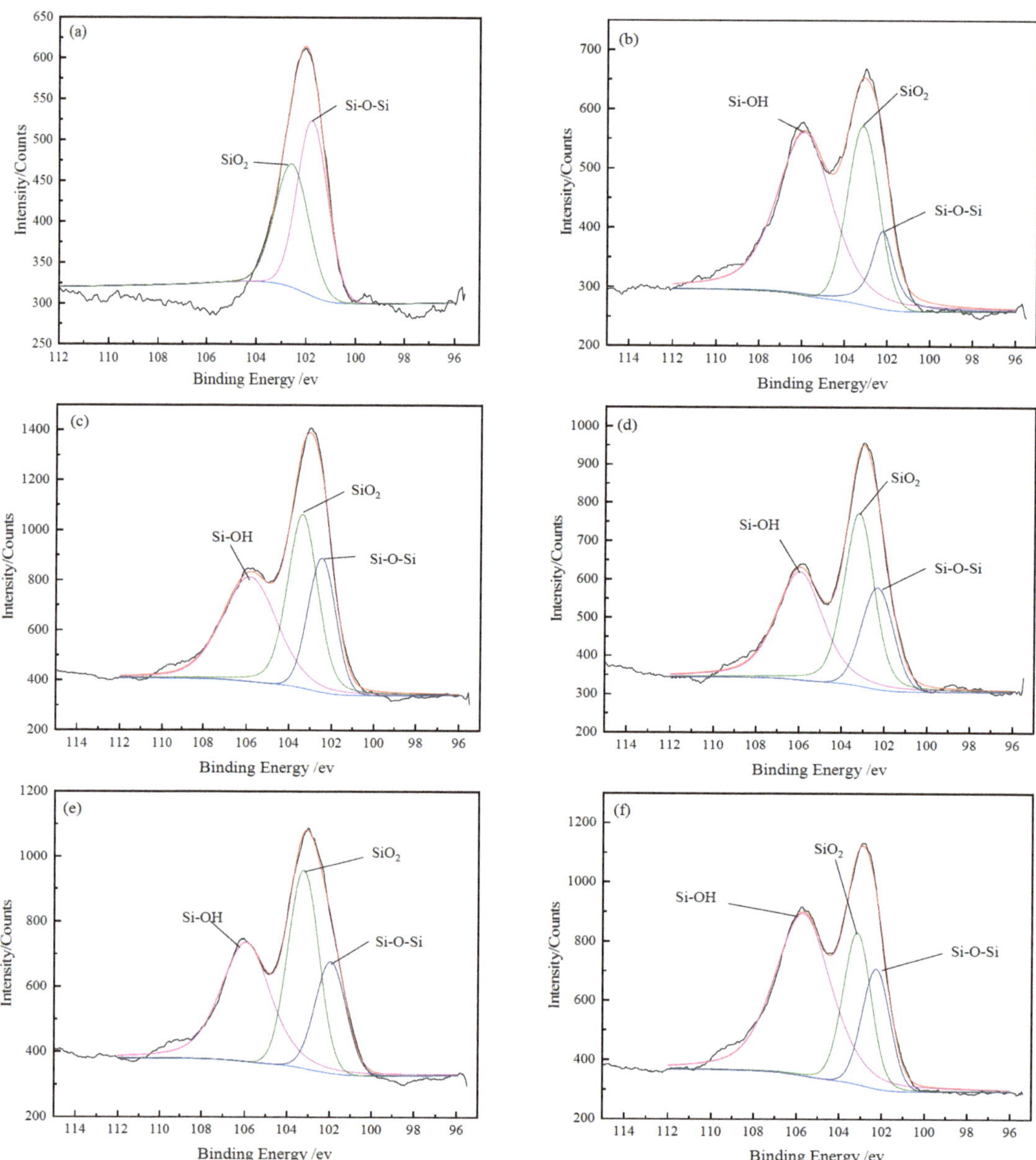

Figure 5. Fitted curves of Si element spectra: (**a**) untreated; (**b**) treated with modification solution containing La^{3+} 0.1 wt.%; (**c**) treated with modification solution containing La^{3+} 0.3 wt.%; (**d**) treated with modification solution containing La^{3+} 0.5 wt.%; (**e**) treated with modification solution containing La^{3+} 0.7 wt.%; (**f**) treated with modification solution containing La^{3+} 0.9 wt.%.

It can be seen that the SiO_2 content on the BFs surfaces firstly decreased and then increased with La^{3+} concentration increasing, but the content changes of Si–O–Si and Si–OH were contrary. When La^{3+} concentration was 0.5 wt.%, the SiO_2 content reached a minimum value (36.12%) corresponding to the highest contents of Si–O–Si and Si–OH. With the La^{3+} concentration beyond 0.5 wt.%, the ether bond was generated with the reaction of Si–OH and active groups in the rare earth modification solution. Accordingly, the hydrolysis of SiO_2 was compressed and the content of SiO_2 increased.

3.3. Basalt Fiber Surface Morphology

Figure 6 reveals the changes of BFs surface morphology after modification. The surfaces of untreated BFs were smooth without obvious grooves or salients in Figure 6a. Some particles formed through the process of the active groups drawn onto the fiber surfaces by the rare earth element, were attached to the surfaces of BFs treated with the rare earth modification solution containing La^{3+} concentration 0.1 wt.% in Figure 6b. Figure 6c,d shows that a trend of increasing particles was accompanied by an increasing concentration of La^{3+}. When La^{3+} concentration was up to 0.5 wt.%, a great number of small particles uniformly covered the fiber surfaces to form the fibers' coatings.

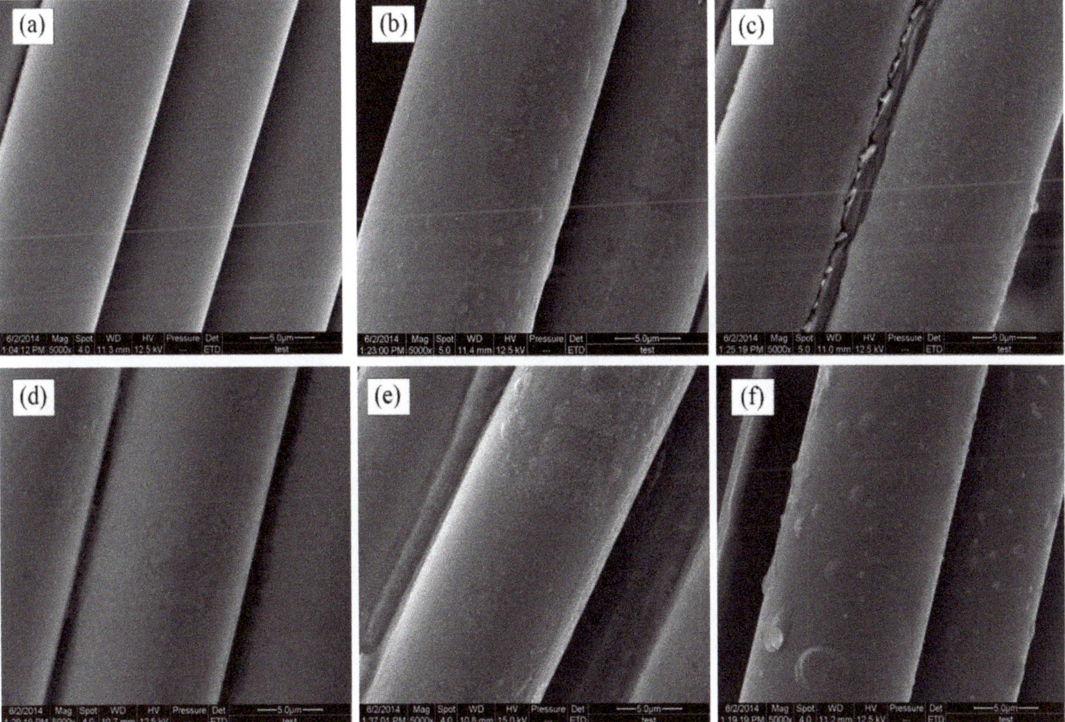

Figure 6. SEM images of BFs: (**a**) untreated; (**b**) treated with modification solution containing La^{3+} 0.1 wt.%; (**c**) treated with modification solution containing La^{3+} 0.3 wt.%; (**d**) treated with modification solution containing La^{3+} 0.5 wt.%; (**e**) treated with modification solution containing La^{3+} 0.7 wt.%; (**f**) treated with modification solution containing La^{3+} 0.9 wt.%.

As shown in Figure 6e,f, the particles decreased on the fiber surfaces, as La^{3+} concentration increased to 0.7 wt.%. When La^{3+} concentration was up to 0.9 wt.%, the excess rare earth elements were attached onto the fiber surfaces with the saturation of active groups, resulting that only a few particles of inhomogeneous size were on the fibers. Moreover, these active groups were easy to bulge and even be broken under the function of tension.

3.4. Mechanical Properties of Basalt Fibers/Epoxy Resin Composites

Tensile, bending and ILSS standard tests were performed to analyze mechanical behavior of BF/ERCs. According to the obtained experimental data shown in Figures 7–9, the mechanical properties, including the tensile strength, the bending strength and ILSS of BF/ERCs, were improved by BFs treated with La^{3+} concentration increasing from 0.1 to 0.5 wt.%. The active functional groups attached to the fibers' surfaces, such as C=O and O–H, improved the fiber surface roughness and activity, and thus enhanced the adhesion between fibers and the composite matrix. The maximum values of the tensile strength, the bending strength and ILSS were up to 458.7, 556.7 and 16.77 Mpa, and increased by 56.22%, 103.32% and 88% compared with those of the unmodified ones, respectively.

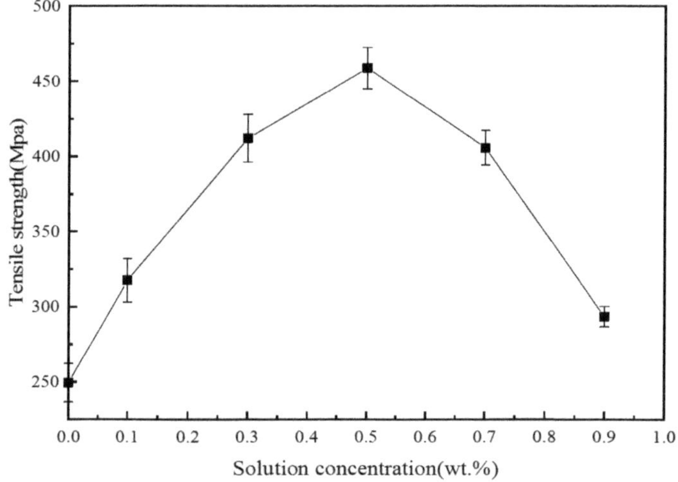

Figure 7. Tensile strength of BF/ERCs.

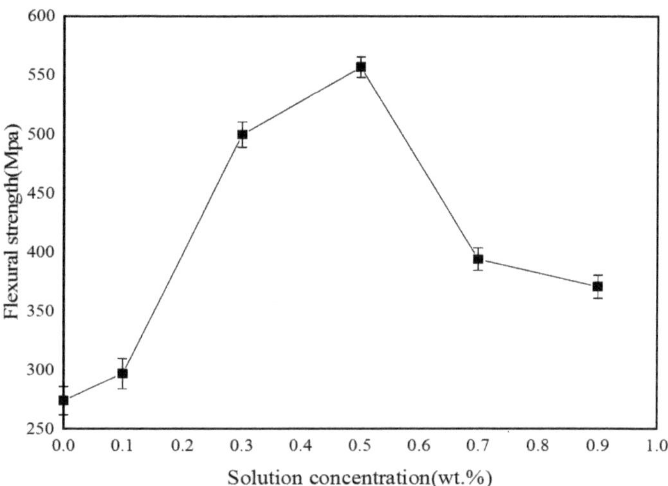

Figure 8. Flexural strength of BF/ERCs.

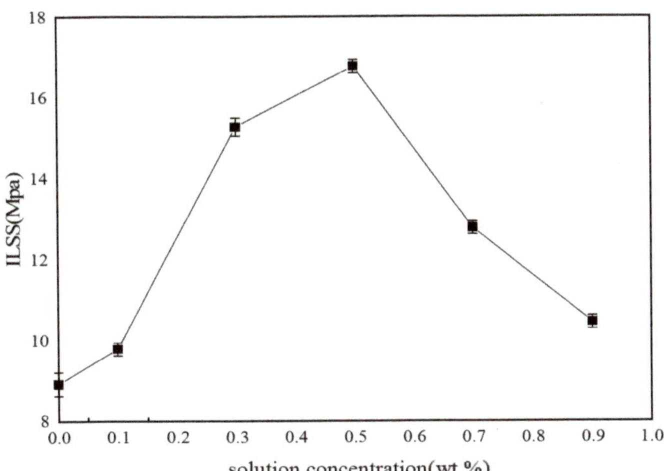

Figure 9. ILSS of BF/ERCs.

When La^{3+} concentration exceeded 0.5 wt.%, the mechanical strength of composites decreased as La^{3+} concentration increased. This may be ascribed to the fact that the excess active groups introduced by La^{3+} were aggregated to be supersaturated, and the triggered break of saturated active groups happened, leading to the loss of active groups. Moreover, the bits of grains on the interface of composite formed by the introduction of the excess La^{3+} to the fibers' surface also reduced the interface bonding strength.

3.5. Fracture Surfaces Morphology of Basalt Fibers/Epoxy Resin Composites

The fracture surfaces of untreated and treated BF/ERCs in the tensile test was observed by SEM. In Figure 10a, smooth fibers and grooves can be seen on the fracture surfaces of untreated composites. Caused by the worse interfacial adhesion between fibers and the resin matrix, the fibers can be easily separated or pulled out from the resin matrix, exhibiting the low tensile strength of composites.

There were a few resins on the fiber surfaces of composites treated with the rare earth modification solution containing La^{3+} concentration 0.1 wt.%, as shown in Figure 10b. Resins increased and grooves decreased on the fracture surfaces with the increase of La^{3+} concentration, as seen in Figure 10c,d. These residual resins on the fibers' surfaces were attributed in the improved adhesion between fibers and the resin matrix, when fibers were stretched from the resin matrix under the stress. When La^{3+} concentration was up to 0.5 wt.%, the gaps between fibers were filled with the resins in Figure 10d. BFs and the resin matrix adhered so tightly that external load was delivered to the fibers from the matrix, and the fibers mainly tolerated the breaking stress. It exhibited the obviously improved mechanical properties of BF/ERCs in the better reinforcing effect.

When La^{3+} concentration exceeded 0.5 wt.%, the decrease in resins on the fiber surfaces with increasing La^{3+} concentration led to the reappeared grooves on the resin matrix, when the smooth fibers pulled out, as seen in Figure 10e,f. It demonstrated that the interfacial strength between the fibers and the resin matrix reduced, and the mechanical properties of composites correspondingly decreased.

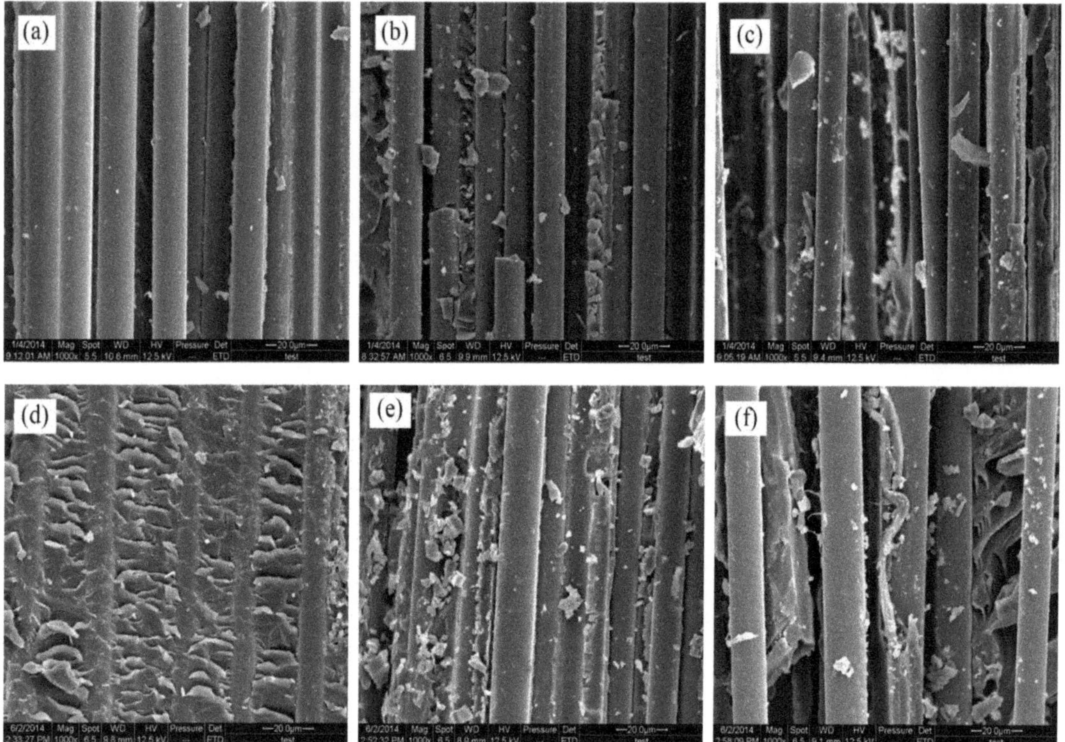

Figure 10. SEM images of fracture surfaces of BF/ERCs: (**a**) untreated; (**b**) treated with modification solution containing La^{3+} 0.1 wt.%; (**c**) treated with modification solution containing La^{3+} 0.3 wt.%; (**d**) treated with modification solution containing La^{3+} 0.5 wt.%; (**e**) treated with modification solution containing La^{3+} 0.7 wt.%; (**f**) treated with modification solution containing La^{3+} 0.9 wt.%.

The monomolecular layer theory can be supplemented to analyze the effect of element La concentration on the tensile performances of BF/ERCs with the model shown in Figure 11. At low concentrations, few La^{3+} were absorbed onto the BFs surfaces. As shown in Figure 11a, the arrangement of La atoms on BFs surfaces was discontinuous with a lot of voids, resulting in the failure of BF/ERCs occurring firstly on the location without La atoms under external load. Thus, the discontinuous interface between BFs and the epoxy resin is insufficient to improve the mechanical performance of BF/ERCs obviously. When La element increased to the best concentration, 0.5 wt.%, the uniform and compact monomolecular layer on the BFs surfaces exhibits high adhesive strength between BFs and the matrix, as shown in Figure 11b. The mechanical performances of BF/ERCs reached to the highest value, including the tensile strength, the bending strength, and ILSS.

When La^{3+} concentration exceeded 0.5 wt.%, excessive La atoms were assembled on the BFs surfaces to form multimolecular layer, as shown in Figure 12. Under external load, failure of BF/ERCs firstly occurred on the interlayer of La atoms interlinked by weak Van der Waals force, resulting in the decreased performance of BF/ERCs.

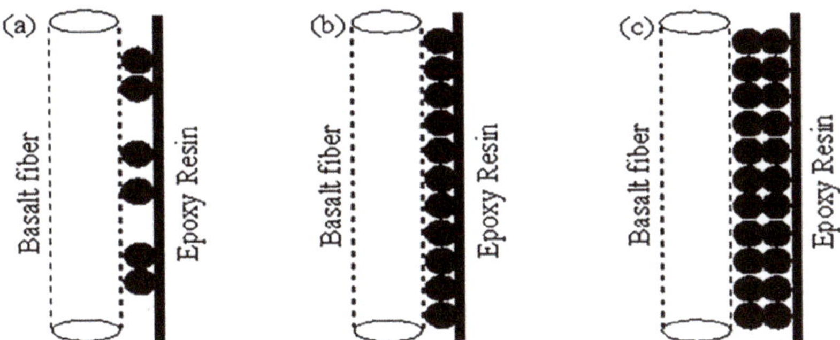

Figure 11. Model of monomolecular layer theory. (**a**) the model with less La^{3+}; (**b**) the model with enough La^{3+}; (**c**) the model with excess La^{3+}.

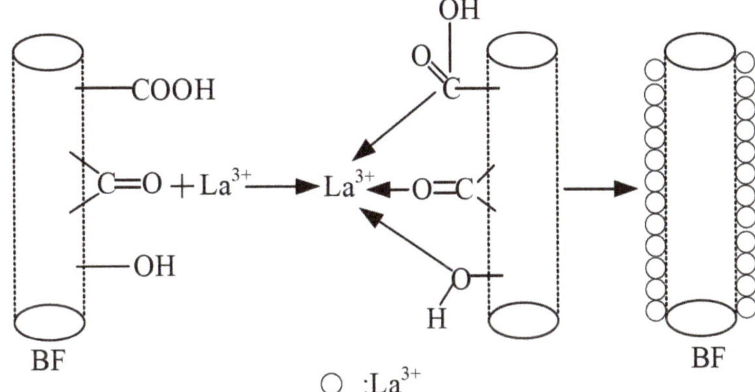

Figure 12. La^{3+} attached onto BFs surfaces.

4. Discussion

4.1. Analysis about Modification Mechanism of La^{3+}

In Figures 1–5, the analysis based on the FT-IR spectra and the XPS results of BFs shows that, compared to untreated BFs, active oxygen-containing functional groups on the fiber surfaces increased, once BFs is treated with the rare earth modification solution. That was ascribed to the mechanism of La element in the modification solution as follows.

The rare earth elements with 4f electronic shell have good chemical activity. Once polarized, they have strong affinity to the nonmetal elements and turn into active elements in alcohol solution with the typical nonmetal elements such as C, H, O and N [26–28]. Consequently, with the results in Section 3.2.2., the coordination bonds between La ions and nonmetal elements are formed and La elements are attached onto the fiber surfaces, as shown in Figure 12.

The coordination number of La element changes from 3 to 12, while the majority level remains 8. In this study, the rare earth modification solution was the alcohol solution with a variety of organic compounds, such as urea and citric acid, as shown in Figure 13a. As the active chemical cores, La ions attached on the BFs surfaces were combined with a lot of active organic groups in the solution to generate the multicomponent complex [26–28], shown in Figure 13b. With the active organic groups absorbed onto the fiber surfaces to form the macromolecular membrane, the chemical activity on the fiber surfaces could be improved obviously, as shown in Figure 13c. As seen in Figure 6, the La ions can also be embedded in the defect points on the BFs surfaces to generate more active chemical cores, and hence the chemical activity of the fiber surfaces can be further improved.

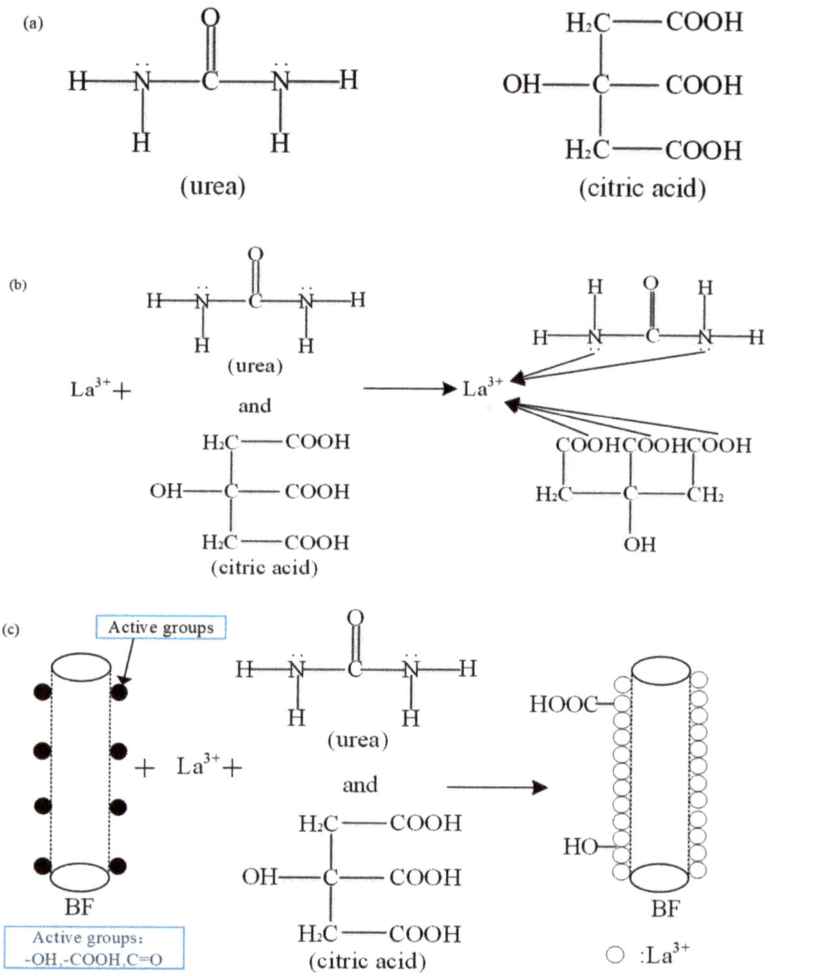

Figure 13. The chemical reaction mechanism of the La^{3+} modification solution: (**a**) the molecular formula of urea and citric acid in the modification solution; (**b**) the chemical combination between La^{3+} and the active groups in the modification solution; (**c**) the chemical combination between the BFs and the modification to introduce the more active groups.

However, the abundant active groups introduced by the excess La^{3+} were aggerated on the fibers' surfaces to be supersaturated, resulting in the break of the macromolecular membrane and loss of active groups, as shown in Figure 4e,f. Meanwhile, with the excess La^{3+}, the generated Si–OH between SiO_2 of the fibers' surface and the modification solution under hydrolytic action restrain the further hydrolytic action of SiO_2, leading to the decrease of Si–OH and Si–O–Si in Figure 4e,f, which further reduced the surface activity of fibers. Thus, the La^{3+} concentration of 0.5% was the most suitable.

4.2. Analysis of Enhancement Mechanism of BFs/ECR's Mechanical Property Modified with La^{3+}

In the process of the resin matrix synthesis, the following crosslinking reaction of epoxy resin with ali-phatic polyamine curing agent took place and generated two oxhydryl groups, as shown in Figure 14.

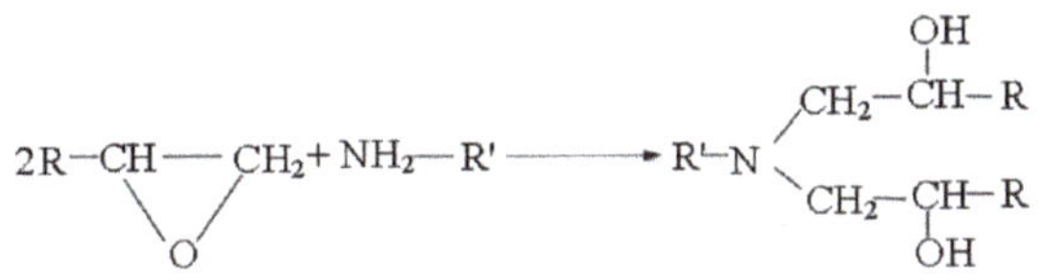

Figure 14. The crosslinking reaction of epoxy resin with aliphatic polyamine curing agent.

Based on the FT-IR spectra and the XPS results of BFs, the active functional groups increased on the surface of BFs treated with the rare earth modification solution. With the oxhydryl generated in the above crosslinking reaction, the active functional groups took part in the dehydration condensation reaction, when BFs were in sufficient contact with epoxy resin. The bonding of BFs and epoxy resin was based on the chemical bonds, such as the ester group and the ether bond, so that the bonding force improved significantly, as shown in Figure 15a. Thus, when La^{3+} concentration was 0.5 wt.%, the absorbed active groups reached to the maximumand the bonding force was correspondingly highest, as well as the tensile strength, the bending strength, and ILSS.

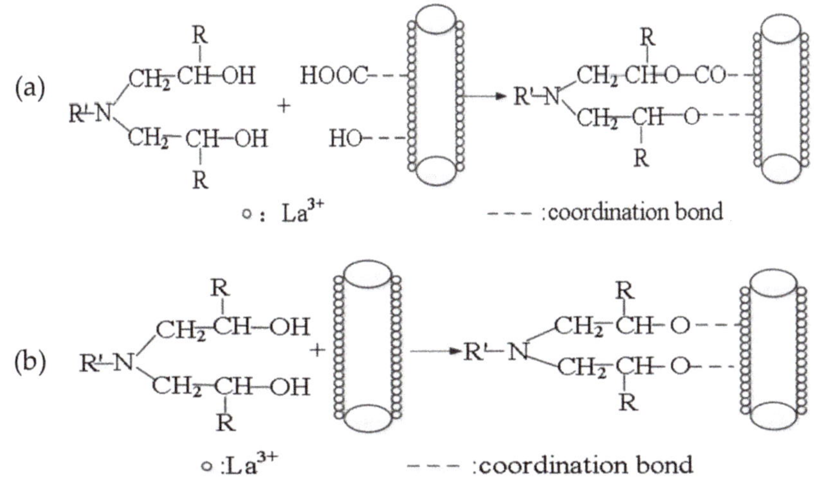

Figure 15. Effect of La^{3+} on the interface of BFs and epoxy resin. (**a**) the chemical reaction between the absorbed active groups and the epoxy resin containing curing agent; (**b**) the coordination reaction between La^{3+} and the epoxy resin containing curing agent.

However, parts of La ions, which were absorbed on the modified BFs surfaces, did not take part in the coordination bond with active groups. When the modified BFs were put into the epoxy resin, these La ions can be combined with oxhydryl generated in the above crosslinking reaction, because of the typical chemical activities and the variable coordination number of La^{3+}. As shown in Figure 15b, under the effect of the formed coordination bond, the adhesion strength between BFs and epoxy matrix was further improved, as well as the mechanical performances of BF/ERCs.

5. Conclusions

In this research, the modification of basalt fibers with the modification solution of different La^{3+} concentrations was carried out. Moreover, the effect of modification with different La^{3+} concentrations and the mechanical properties for basalt fiber/epoxy resin composites modified with La^{3+} were evaluated.

The study about the chemical composition, the functional groups and the morphology of untreated and treated BFs' surface with the rare earth modification solution using several analytical techniques (FTIR, XPS, and SEM), confirmed that La element in the rare earth modification solution could link active oxygen-containing functional groups to the BFs surfaces to improve the roughness and the activity of the fiber surfaces. When the La^{3+} concentration in the rare earth modification solution was 0.5 wt.%, the activity of BFs' surface reached the strongest with the most introduced active groups. It is also determined that the mechanical performances of BF/ERCs, including the tensile strength, the bending strength and ILSS, could be significantly enhanced by the modified BFs due to the formed chemical reaction between resin matrix and the more active groups introduced by La^{3+}. With the La^{3+} concentration in the rare earth modification solution reaching 0.5 wt.%, the bonding between the resin matrix and BFs is demonstrated to be the best and the tensile strength, the bending strength and ILSS were up to 458.7, 556.7 and 16.77 Mpa, and improved by 56.22%, 103.32% and 88% compared with those of the unmodified ones, respectively.

In addition, the modification mechanism of La^{3+} and enhancement mechanism of BFs/ECR's mechanical property modified with La^{3+} were stated with the strong affinity of La to the nonmetal elements, indicating that the modification of BFs with La^{3+} was effective and the mechanical property of the prepared BFs/ECR composites was excellent.

Author Contributions: Conceptualization, C.L.; methodology, C.L.; software, H.W.; validation, C.L., X.H. and Y.F.; formal analysis, C.L.; investigation, C.L., H.W. and Y.S.; resources, C.L.; data curation, Y.S.; writing—original draft preparation, C.L. and H.W.; writing—review and editing, C.L., H.W. and X.Z.; visualization, H.W. and X.Z.; supervision, C.L.; project administration, C.L.; funding acquisition, C.L. All authors have read and agreed to the published version of the manuscript.

Funding: The research was funded by Science Foundation of the National Key Laboratory of Science and Technology on Advanced Composites in Special Environment (Grand No. JCKYS2019603C006).

Institutional Review Board Statement: Not applicable.

Informed Consent Statement: Not applicable.

Data Availability Statement: The data presented in this study are available on request from the corresponding author. The data are not publicly available as the data also form part of an ongoing study.

Conflicts of Interest: The authors declare no conflict of interest.

References

1. Wei, B.; Song, S.; Cao, H. Strengthening of basalt fibers with nano-SiO$_2$–epoxy composite coating. *Mater. Des.* **2011**, *32*, 4180–4186. [CrossRef]
2. Dhand, V.; Mittal, G.; Rhee, K.Y.; Park, S.J.; Hui, D. A short review on basalt fiber reinforced polymer composites. *Compos. Part B Eng.* **2015**, *73*, 166–180. [CrossRef]
3. Manikandan, V.; Jappes, J.; Kumar, S.; Amuthakkannan, P. Investigation of the effect of surface modifications on the mechanical properties of basalt fibre reinforced polymer composites. *Compos. Part B Eng.* **2012**, *43*, 812–818. [CrossRef]
4. Lopresto, V.; Leone, C.; De Iorio, I. Mechanical characterisation of basalt fibre reinforced plastic. *Compos. Part B Eng.* **2011**, *42*, 717–723. [CrossRef]
5. Colombo, C.; Vergani, L.; Burman, M. Static and fatigue characterisation of new basalt fibre reinforced composites. *Compos. Struct.* **2012**, *94*, 1165–1174. [CrossRef]
6. Wei, B.; Cao, H.; Song, S. Surface modification and characterization of basalt fibers with hybrid sizings. *Compos. Part A* **2011**, *42*, 22–29. [CrossRef]
7. Scalici, T.; Valenza, A.; Di Bella, G.; Fiore, V. A review on basalt fibre and its composites. *Compos. Part B Eng.* **2015**, *74*, 74–94.
8. Arvin, A.C.; Bakis, C.E. Optimal design of press-fitted filament wound composite flywheel rotors. *Compos. Struct.* **2006**, *72*, 47–57. [CrossRef]

9. Rashkovan, I.A.; Korabel'nikov, Y.G. The effect of fiber surface treatment on its strength and adhesion to the matrix. *Compos. Sci. Technol.* **1997**, *57*, 1017–1022. [CrossRef]

10. Lee, S.O.; Rhee, K.Y.; Park, S.J. Influence of chemical surface treatment of basalt fibers on interlaminar shear strength and fracture toughness of epoxy-based composites. *J. Ind. Eng. Chem.* **2015**, *32*, 153–156. [CrossRef]

11. Kim, M.T.; Kim, M.H.; Rhee, K.Y.; Park, S.J. Study on an oxygen plasma treatment of a basalt fiber and its effect on the interlaminar fracture property of basalt/epoxy woven composites. *Compos. Part B Eng.* **2011**, *42*, 499–504. [CrossRef]

12. Schneck, T.K.; Brück, B.; Schulz, M.; Spörl, J.; Hermanutz, F.; Clauß, B.; Mueller, W.M.; Heidenreich, B.; Koch, D.; Horn, S.J.; et al. Carbon fiber surface modification for tailored fiber-matrix adhesion in the manufacture of C/C-SiC composites. *Compos. Part A Appl. Sci. Manuf.* **2019**, *120*, 64–72. [CrossRef]

13. Jiang, T.; Li, Z.; Xiao, P.; Liu, J.; Liu, P.; Yu, S.; Xiao, T.; Liu, L.; Technology, C. Effect of different chemical treatments on hydroxyapatite formation of carbon fibers reinforced carbon and SiC dual matrices composites. *Surf. Coat. Technol.* **2019**, *357*, 153–160. [CrossRef]

14. Dighton, C.; Rezai, A.; Ogin, S.L.; Watts, J.F. Atmospheric plasma treatment of CFRP composites to enhance structural bonding investigated using surface analytical techniques. *Int. J. Adhes. Adhes.* **2019**, *91*, 142–149. [CrossRef]

15. Cho, B.G.; Hwang, S.H.; Park, M.; Park, J.K.; Park, Y.B.; Chae, H.G.J.C. The effects of plasma surface treatment on the mechanical properties of polycarbonate/carbon nanotube/carbon fiber composites. *Compos. Part B Eng.* **2019**, *160*, 436–445. [CrossRef]

16. Cech, V.; Knob, A.; Lasota, T.; Lukes, J.; Drzal, L.T.J.P.C. Surface modification of glass fibers by oxidized plasma coatings to improve interfacial shear strength in GF/polyester composites. *Polym. Compos.* **2019**, *40*, E186–E193. [CrossRef]

17. Arslan, C.; Dogan, M.J. The effects of fiber silane modification on the mechanical performance of chopped basalt fiber/ABS composites. *J. Thermoplast. Compos. Mater.* **2019**, *33*, 089270571982951. [CrossRef]

18. Latif, R.; Wakeel, S.; Khan, N.Z.; Siddiquee, A.N.; Verma, S.L.; Khan, Z.A. Surface treatments of plant fibers and their effects on mechanical properties of fiber-reinforced composites: A review. *J. Reinf. Plast. Compos.* **2019**, *38*, 15–30. [CrossRef]

19. Iorio, M.; Santarelli, M.L.; González-Gaitano, G.; González-Benito, J. Surface modification and characterization of basalt fibers as potential reinforcement of concretes. *Appl. Surf. Sci.* **2018**, *427*, 1248–1256. [CrossRef]

20. Xu, Z.; Wu, X.; Ying, S.; Jiao, Y.; Li, J.; Li, C.; Lu, L. Surface modification of carbon fiber by redox-induced graft polymerization of acrylic acid. *J. Appl. Polym. Ence.* **2010**, *108*, 1887–1892. [CrossRef]

21. Jiang, Z.X.; Liu, L.; Huang, Y.D.; Ren, H. Influence of coupling agent chain lengths on interfacial performances of carbon fiber and polyarylacetylene resin composites. *Surf. Interface Anal.* **2010**, *41*, 624–631. [CrossRef]

22. Li, K.Z.; Wang, C.; Li, H.J.; Li, X.T.; Ouyang, H.B.; Jian, W. Effect of chemical vapor deposition treatment of carbon fibers on the reflectivity of carbon fiber-reinforced cement-based composites. *Compos. Sci. Technol.* **2008**, *68*, 1105–1114. [CrossRef]

23. Hung, K.B.; Li, J.; Fan, Q.; Chen, Z.H. The enhancement of carbon fiber modified with electropolymer coating to the mechanical properties of epoxy resin composites. *Compos. Part A Appl. Sci. Manuf.* **2008**, *39*, 1133–1140. [CrossRef]

24. Jain, N.; Singh, V.K.; Chauhan, S. Review on effect of chemical, thermal, additive treatment on mechanical properties of basalt fiber and their composites. *J. Mech. Behav. Mater.* **2018**, *26*, 205–211. [CrossRef]

25. Ricciardi, M.R.; Papa, I.; Coppola, G.; Lopresto, V.; Sansone, L.; Antonucci, V. Effect of plasma treatment on the impact behavior of epoxy/basalt fiber-reinforced composites: A preliminary study. *Polymers* **2021**, *13*, 1293. [CrossRef] [PubMed]

26. Ju, W.; Cheng, X. Effect of rare earth modifier treatment on interlaminar shear strength of aramid/epoxy composites. *Rare Met. Mater. Eng.* **2005**, *34*, 1917–1920.

27. Bao, D.; Cheng, X.H. The effect of rare earth treatment on tensile properties of carbon fiber reinforced polytetrafluoroethylene composites. *J. Shanghai Jiaotong Univ.* **2006**, *40*, 914–917.

28. Wu, J.; Cheng, X.H. Effect of rare earth treatment on tensile properties of F-12 fiber reinforced epoxy composites. *J. Shanghai Jiaotong Univ.* **2005**, *39*, 1795–1798.

29. Jin, T.; Shen, S.; Jing, L.I.; Weina, L.I.M.R. Feasibility study of a new method of basalt fiber surface treatment-acid etching treatment. *Mater. Rev.* **2014**, *12*, 116–118.

Article

Fabrication and Properties of Electrospun and Electrosprayed Polyethylene Glycol/Polylactic Acid (PEG/PLA) Films

Weichang Ke [1], Xiang Li [2], Mengyu Miao [2], Bing Liu [1], Xiaoyu Zhang [2] and Tong Liu [2],*

1 China Tobacco Hubei Industrial LLC, Wuhan 430040, China; kewch@hbtobacco.cn (W.K.); jerry.lb@163.com (B.L.)
2 School of Power and Mechanical Engineering, Wuhan University, Wuhan 430072, China; 2017302650114@whu.edu.cn (X.L.); 2015302650129@whu.edu.cn (M.M.); 2016302650148@whu.edu.cn (X.Z.)
* Correspondence: liu_tong@whu.edu.cn

Abstract: Polylactic acid (PLA) film is an alternative filter material for heat-not-burn (HNB) tobacco, but its controllability in cooling performance is limited. In this work, polyethylene glycol (PEG) was introduced to form a polyethylene glycol/polylactic acid (PEG/PLA) film by electrospinning or electrospraying techniques to enhance the cooling performance, due to its lower glass transition and melting temperatures. The PEG/PLA films with typical electrospun or electrosprayed morphologies were successfully fabricated. One typical endothermic peak at approximately 65 °C was clearly observed for the melting PEG phase in the heating process, and the re-crystallization temperature represented by an exothermic peak was effectively lowered to 90–110 °C during the cooling process, indicating that the cooling performance is greatly enhanced by the introduction of the PEG phase. Additionally, the wetting properties and adsorption properties were also intensively studied by characterizing the contact angles, and the as-prepared PEG/PLA films all showed good affinity to water, 1,2-propandiol and triglyceride. Furthermore, the PEG/PLA film with a PLA content of 35 wt.% revealed the largest elasticity modulus of 378.3 ± 68.5 MPa and tensile strength of 10.5 ± 1.1 MPa. The results achieved in this study can guide the development of other filter materials for HNB tobacco application.

Keywords: polylactic acid; polyethylene glycol; electrospinning; heat-not-burn tobacco; cooling performance

Citation: Ke, W.; Li, X.; Miao, M.; Liu, B.; Zhang, X.; Liu, T. Fabrication and Properties of Electrospun and Electrosprayed Polyethylene Glycol/Polylactic Acid (PEG/PLA) Films. *Coatings* **2021**, *11*, 790. https://doi.org/10.3390/coatings11070790

Academic Editors: Csaba Balázsi and Kevin Plucknett

Received: 10 June 2021
Accepted: 28 June 2021
Published: 30 June 2021

Publisher's Note: MDPI stays neutral with regard to jurisdictional claims in published maps and institutional affiliations.

Copyright: © 2021 by the authors. Licensee MDPI, Basel, Switzerland. This article is an open access article distributed under the terms and conditions of the Creative Commons Attribution (CC BY) license (https://creativecommons.org/licenses/by/4.0/).

1. Introduction

Smoking has an extremely long history and is deeply rooted in society. The total consumption of conventional cigarettes, the most popular tobacco product, reached 5.7 trillion in 2016 [1]. As a result, smoked cigarette butts have become one of largest contributors of litter in the world. It has been largely reported that smoked cigarette butts contain over 7000 chemicals, including many harmful and/or carcinogenic effects, which can be released or leached after being released into the environment [2–4]. Therefore, new cigarette types, such as heat-not-burn (HNB) tobacco and electronic cigarettes, have been developed to overcome these threats to life and to the environment. HNB tobacco devices heat tobacco but do not burn it, which can reduce the release of harmful products from its combustion [5–7]. Bentley et al. reported that the emission of toxic products can be reduced by approximately 95% [8]. However, harmful gases, including formaldehyde, acetaldehyde and acrolein, may still be generated in mainstream smoke due to the use of propylene glycol and vegetable glycerin in HNB tobacco [9,10].

Recently, adsorbing materials, such as zeolites, metal–organic frameworks, activated carbon, fibers, coordination polymers and carbon nanotubes, have been developed to further remove the toxic chemicals in mainstream smoke due to their excellent adsorbing properties provided by their extremely high specific surface areas [11–13]. However, there are still many limitations in processing and production, such as the material safety, high cost and tedious preparation process. Therefore, it is critical to design and manufacture novel, more efficient adsorbing materials for HNB tobacco products.

Biodegradable natural polymer polylactic acid (PLA) is an abundant renewable biomass polymer, which has been applied as an HNB filter material due to its low cost, lightweight and apparently environmentally superiority alternative to synthetic materials [14–16]. It is proven that the harmful pollutants to the environment during the smoking process have been greatly eliminated by the application of PLA filters for HNB tobacco, thus presenting strongly positive effects on the economy, human beings and society [17]. However, the cooling performance can only be narrowly tailored by the use of single-phase PLA. Moreover, the controllability in cooling properties is limited.

Currently, the introduction of polymer materials with lower glass transition and melting temperatures to form blending materials has been considered as one of the most effective strategies to manipulate the cooling performance as the as-prepared blending materials effectively exhibit lowered thermal properties, including glass transition temperature, melting temperature and recrystallization temperature [18–23]. Polyethylene glycol (PEG) material, which has a much lower glass transition temperature (below 40 °C) and melting temperature (lower than 70 °C) than those for classic PLA filter materials, is one of the most commonly used polymer additives to adjust cooling properties due to its biodegradability, good cooling performance, easy processing and low cost [18–21]. Therefore, PLA and PEG materials are chosen as the main material and polymer additive to form polyethylene glycol/polylactic acid (PEG/PLA) blending film, respectively, and the cooling performance of PEG/PLA blending material can be enhanced by the glass transition and melting of PEG phase at a lower temperature than PLA material, thus leading to a reduced outlet temperature of the filter of HNB tobacco.

The objective of this work is to prepare novel films with high cooling performance, good wetting and adsorption properties, as well as considerable mechanical properties for HNB tobacco application. The electrospinning technique has been considered as one of the most efficient ways to fabricate continuous fibers in the submicron to nanometer scale range, and the microstructure of the fibers is highly dependent on electrospinning parameters such as precursor electrospinning solution, operating voltage and the distance from the collector to the needle [23–35]. Therefore, in the present work, three typical electrospun/electrosprayed PEG/PLA films with different morphologies were prepared by using the simple electrospinning or electrospray method and adjusting the PLA content and precursor electrospinning solution. Their morphologies were measured by using scanning electron microscopy (SEM), while the thermal properties were determined by differential scanning calorimetry (DSC) and thermal conductivity meters (TCM). Finally, the wetting properties and mechanical properties were evaluated using the contact angle meter and universal material testing machine, respectively.

2. Materials and Methods

2.1. Materials

Polylactic acid (PLA) with an average molecular weight (Mw) of 80,000 g/mol was purchased from Macklin Biochemical Co. Ltd (Shanghai, China). Polyethylene glycol (PEG) with Mw of 10,000 g/mol was supplied by Aladdin Biochemical Technology Co. Ltd (Shanghai, China). Dimethylacetamide (DMAc), the solvent of the electrospinning solution, was purchased from Sinopharm Chemical Reagent Co., Ltd (Shanghai, China). Note that the chemicals used in this work were all of analytical reagent (AR) grade.

2.2. Preparation and Characterization of the Spinning Solutions

PLA and PEG were dissolved in DMAc solvent and stirred at 50 °C till the homogeneous spinning solutions were obtained, and the composition parameters of three spinning solutions used in this work are listed in Table 1. For simplification, these three spinning solutions were marked as PLA25, PLA35 and PLA45, respectively.

Table 1. Composition parameters of the three spinning solutions used in this work.

Sample	Composition (wt.%)		
	DMAc	PEG	PLA
PLA25	70	5	25
PLA35	60	5	35
PLA45	50	5	45

2.3. Electrospinning Conditions

A commercial electrospinning apparatus (Model SS-1, Beijing Ucalery Technology Development Co. Ltd., Beijing, China) was applied to prepare PEG/PLA fiber films. A grounded aluminum cylinder with an outer diameter of 10.0 cm and a length of 32.0 cm, which was covered by silicon paper, was used as the fiber collector, and its rotation speed was fixed at 30 rpm. During the electrospinning process, the distance from the collector to the needle was fixed at 12.0 cm. Meanwhile, the flow rate of the solution was precisely controlled at 2 mL/h by using a syringe pump, while the working temperature and the relative humidity in the compartment were controlled at approximately 40 °C and 35% relative humidity (RH). The high voltage applied in the electrospinning/electrospray process was +15 kV. The PEG/PLA fiber films obtained by electrospinning for more than 3 h were dried at 50 °C in a vacuum oven for 5 h to completely remove the residual solvent. The obtained PEG/PLA films were named PLA25, PLA35 and PLA45, respectively.

2.4. Characterization

The morphology of the PEG/PLA films was examined by using scanning electron microscopy (SEM; MIRA 3, Tescan, Brno, Czech), which was then further analyzed by using the ImageJ software (ImageJ bundled with 64-bit Java 1.8.0_172, National Institutes of Health, Bethesda, MD, USA) to calculate the diameters of the fibers.

The Fourier transform infrared (FTIR) spectra of these films were collected by using an FTIR spectrometer (Nicolet 5700, Thermo Fisher, Waltham, MA, USA).

The thermal characteristics of these films were recorded by differential scanning calorimetry (DSC, Q2000, TA Instruments, New Castle, DE, USA) in the temperature range from 30 to 200 °C under pure N_2 conditions with a flow rate of 30 mL/min. For each sample, a 10 mg PEG/PLA film was placed in an aluminum crucible and then sealed hermetically during a heating and cooling cycle with a heating and cooling rate of 5 °C/min. The phases composition of the fiber films was detected by using an X-ray diffractometer (XRD, X'pert Pro, Malvern Panalytical, Worcester, UK) in the 2-theta (2θ) range of 20°–80°, while the three films with a size of 50 × 50 mm were exposed to pure N_2 to measure their corresponding thermal conductivities at 30 °C by using a thermal conductivity meters (TPS 2500S, Hot Disk, Göteborg, Sweden).

The surface wetting properties of the three PEG/PLA films to water, glycerol and 1,2-propanediol were determined by performing contact angle (CA) measurements via a CA meter (JC2000C, POWEREACH, Shanghai, China) to study their adsorption properties. The average CA values were obtained based on at least 5 independent measurements for each sample.

The mechanical properties of these PEG/PLA films were determined using a universal material testing machine (CMT 6103, MTS Systems Corporation, Yokohama, Japan). The stretching rate was 25 mm/min with an initial length between the clamps of 10 mm at 25 °C. All samples with a size of 30 mm (length) × 3 mm (width) × 0.5 mm (thickness) were cut along the collector rotation direction, and at least 5 samples for each film were tested to calculate the average values for the interpretation of the results.

3. Results and Discussion

3.1. Morphology and Phase Composition

The microstructure of the as-prepared PEG/PLA films shown in Figure 1 significantly changed due to varying the precursor electrospinning solution. PLA25 and PLA35 PEG/PLA films exhibited typical morphologies for the electrospinning method [23–35], while the PLA45 PEG/PLA film showed the characteristic electrosprayed microstructure [33–35]. For the PLA25 film derived from solution PLA25 with the lowest PLA content of 25 wt.%, the outerdiameter (O.D.) of the electrospun PEG/PLA fibers is in the range of 0.10–0.19 μm, while the calculated average O.D. value is 0.14 μm (Figure 1A,a). As the PLA content increased to 35 wt.%, the O.D. of the electrospun PEG/PLA fibers strongly increased to 0.33–2.00 μm, with an average value of 0.80 μm (Figure 1B,b). These phenomena can be explained by the increased viscosity of the solutions induced by the decreased DMAc solvent content and increased PLA binder content, which is also consistent with the previously reported literature [33–37]. However, when further increasing the PLA content to 45 wt.%, no fiber was formed, but homogeneous spherical microsize particles were formed by the PLA45 film (Figure 1C,c), which may be related to the solubility in the solvent, where the intrachain interactions may result in the polymer chains coiling on themselves to form spherical microsize particles [33–35].

Figure 2 shows the XRD patterns of three electrospun/electrosprayed PEG/PLA films mentioned above. Two typical diffraction peaks for PEG (PDF No: 50-2158) and PLA (PDF No: 49-2174), as well as an amorphous bump centered at approximately 23°, can be clearly observed in the 2-theta range of 10°–40°, indicating that these PEG/PLA films are a mixture of amorphous phase and crystalline phase, and no secondary phase was formed during the preparation process. In addition, it was found that the signal of the PLA phase became more observable, while the diffraction peaks for the PEG phase became weaker for the three PEG/PLA films, which can be explained by the gradual increase in PLA content.

Figure 3 shows the FTIR infrared absorption spectra of the three PEG/PLA films. It was noticed that with the increasing PLA content in the films, the absorption bands located at approximately 867, 1161 and 3500 cm^{-1}, which can be attributed to the O–H stretch and associated with PEG content [38], receded. Meanwhile, the C=O stretching vibration peak at 1758 cm^{-1} [39], which can be used to reflect PLA content, was enhanced. These results are strongly consistent with the fact that the PLA content is gradually increased, and no obvious reaction occurs during the preparation process.

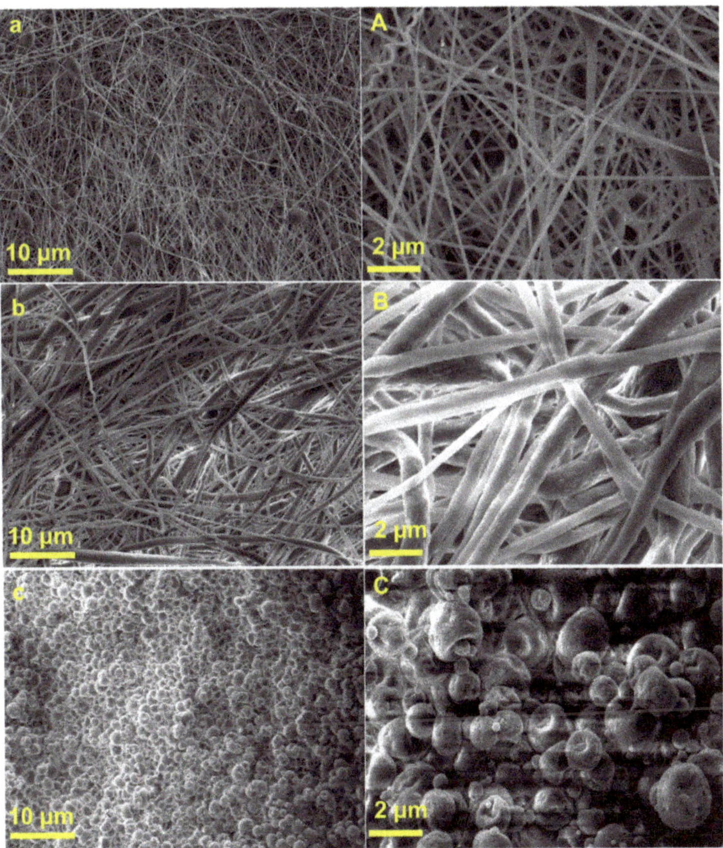

Figure 1. SEM images of (**a**,**A**) PLA25, (**b**,**B**) PLA 35 and (**c**,**C**)PLA45 PEG/PLA films with the magnification of (**a–c**) 5000× and (**A–C**) 20,000×.

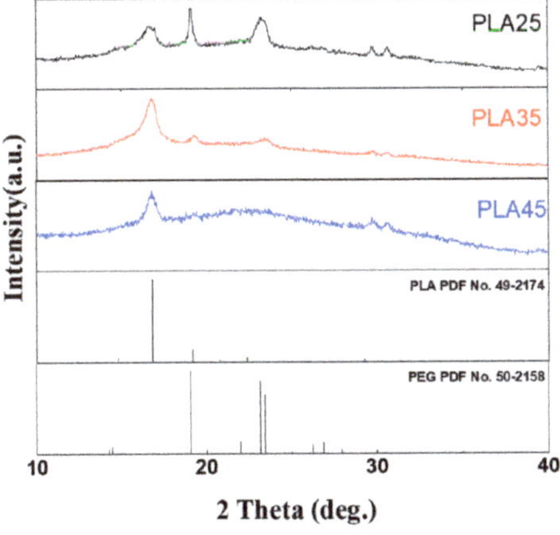

Figure 2. XRD patterns of the three electrospun/electrosprayed PEG/PLA films. (Note that the PDF card numbers for neat PEG and PLA materials are both obtained from MDIJade 6.5 software with the 2004 PDF database.).

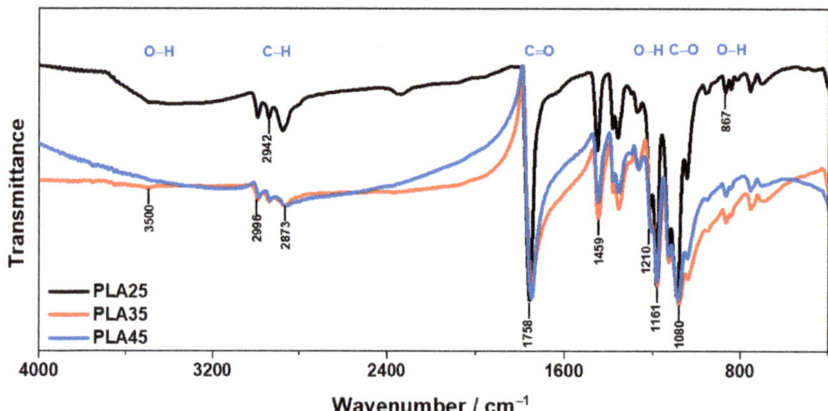

Figure 3. FTIR spectra of the three electrospun/electrosprayed PEG/PLA films.

3.2. Thermal Properties

DSC curves of the three electrospun/electrosprayed PEG/PLA films were measured in the temperature range of 25–200 °C with a cooling and heating rate of 5 °C/min and holding at 200 °C for 30 min to study the thermal transition properties of these samples. Melting temperature, glass transition temperature and crystallization temperature, as well as exothermic and endothermic heat, can be determined in the DSC curves. As shown in Figure 4, in the temperature range of 25–100 °C, during the heating process, typical endothermic peaks at approximately 65 °C could be observed, which are possibly attributed to the glass transition of PLA phase as well as the melting of PEG phase. However, no characteristic peak for the glass transition of PLA phase, a step change in the base line of the scan at approximately 65 °C, was observed. A possible reason for this is that the endothermic peak for the glass transition of PLA phase may be buried due to the stronger melting peak of PEG phase, which can be confirmed by the lowered strength of the endothermic peak with the increasing amount of PLA phase. It should be highlighted that the endothermic peak for the melting of PEG phase can serve as the additional heat absorber during the cooling process, thus reducing the temperature of the flowing gas and improving the cooling performance of PLA material. The characteristic endothermic peak with an onset temperature of approximately 160 °C is ascribed to the melting of PEG/PLA films, which is lower than approximately 170 °C for PLA phase [16]. At the same time, the only typical exothermic peak observed at 90–110 °C during the cooling process is probably associated with the recrystallization of PEG/PLA composite materials, and the temperature is obviously lower than that for pure PLA material [16]. These results, including the low-temperature endothermal peak at 65 °C for melting PEG phase, decreased melting temperature from 170 to 160 °C and lowered onset recrystallization from approximately 130 to 110 °C, indicate that the addition of PEG can effectively improve the cooling performance, and the increase in PEG content can further optimize the cooling performance. Note that a small exothermic peak was observed at approximately 70 °C, which is possibly ascribed to the evaporation of the remaining DMAc solvent in the film.

Figure 5 shows the thermal conductivities of the three electrospun/electrosprayed PEG/PLA films measured at 30 °C. The thermal conductivity effectively increased from 0.1072 W·m⁻¹·K⁻¹ for the PLA25 film to 0.2046 and 0.2237 W·m⁻¹·K⁻¹ for the PLA35 film and the PLA45 film, respectively. These results imply that a higher PLA content and lower porosity in the electrospun/electrosprayed PEG/PLA films can benefit heat transfer and can further enhance the cooling performance.

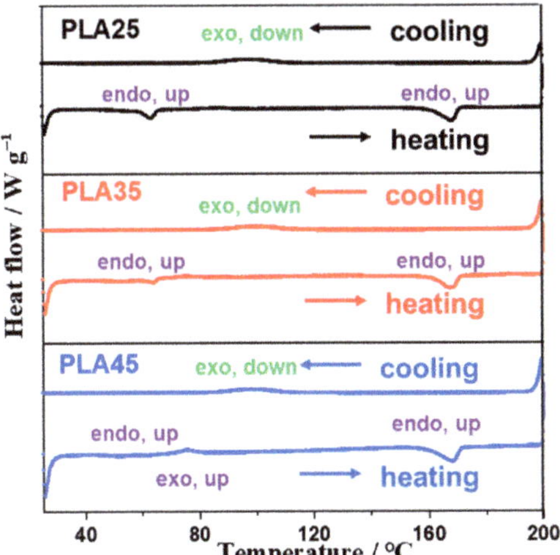

Figure 4. DSC curves of the three electrospun/electrosprayed PEG/PLA films.

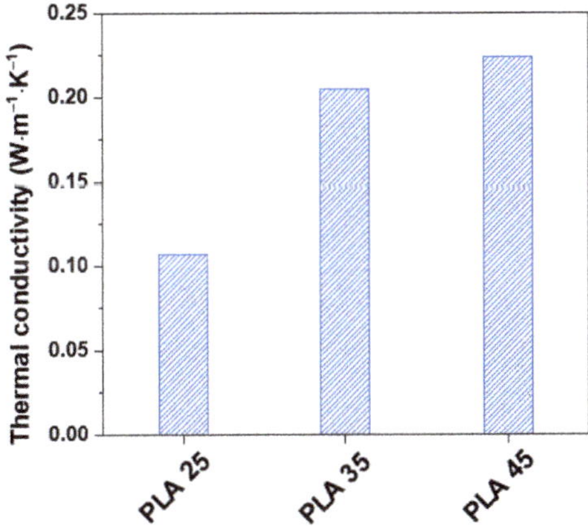

Figure 5. Thermal conductivities of the three electrospun/electrosprayed PEG/PLA films measured at 30 °C.

3.3. Wetting and Adsorption Properties

The wetting properties of the three electrospun/electrosprayed PEG/PLA films towards water, 1,2-propanediol and triglyceride were also studied and are summarized in Figure 6. The contact angles were $88.4° \pm 1.5°$, $34.4° \pm 1.7°$ and $42.3° \pm 1.2°$ when the PLA25 film was exposed to water, 1,2-propandiol and triglyceride, respectively. As the PLA content increased to 35 wt.%, the corresponding contact angle values for the PLA35 film gradually increased to $111.4° \pm 0.8°$, $36.7° \pm 3.6°$ and $51.2° \pm 2.5°$, respectively. However, by further increasing PLA content, the contact angle to water slightly decreased to $100.9° \pm 3.2°$, while the values to 1,2-propanediol and triglyceride continually increased to $43.7° \pm 2.8°$ and $53.0° \pm 3.5°$, respectively. The increase in contact angle value implies the decrease in wetting properties and adsorption properties of the films, which may be explained by the increased PLA content with lower adsorption properties [40].

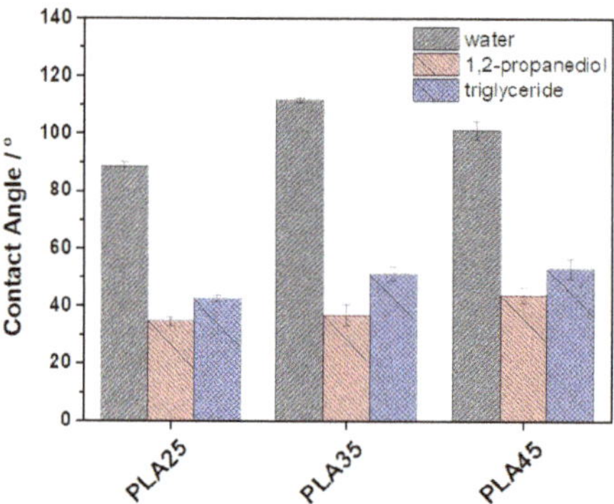

Figure 6. Contact angles for the three electrospun/electrosprayed PEG/PLA films.

3.4. Mechanical Properties

The typical strain–stress curves for three different PEG/PLA films shown in Figure 7 were measured by using the tension method. The mechanical properties, involving maximum elasticity modulus, tensile break stress, tensile strength, tensile yield stress and maximum force, are summarized in Table 2. It was found that the PLA35 film exhibited the largest value with the maximum elasticity modulus of 378.3 ± 68.5 MPa, the tensile break stress of 4.3 ± 0.5 MPa, the tensile strength of 10.5 ± 1.1 MPa, the tensile yield stress of 9.5 ± 1.0 MPa and the maximum force of 0.31 ± 0.03 N, respectively. The PLA45 film showed the peak tensile strain at break of 67.9% ± 21.4%, while the PLA25 film revealed the worst mechanical properties. The mechanical properties are different for these three films, which is possibly attributed to their different porosities and morphologies, requiring further study in the future.

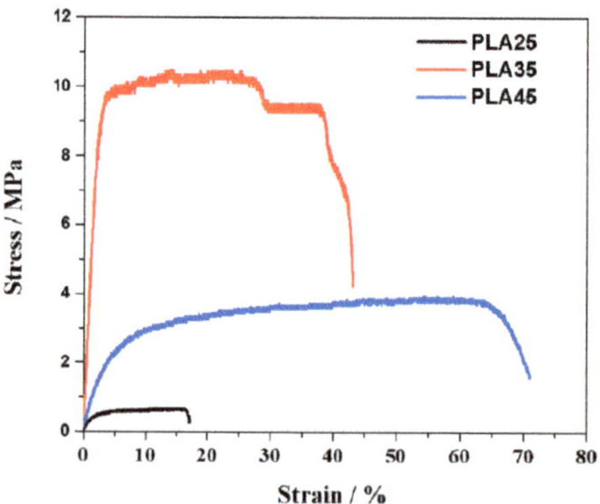

Figure 7. Strain–stress curves for the three electrospun/electrosprayed PEG/PLA films.

Table 2. Summary of the mechanical properties of the three electrospun/electrosprayed PEG/PLA films.

Mechanical Property	PLA25	PLA35	PLA45
Elasticity modulus (MPa)	20.1 ± 4.0	378.3 ± 68.5	50.0 ± 16.4
Tensile strain at break (%)	18.4 ± 3.6	43.0 ± 11.6	67.9 ± 21.4
Tensile break stress (MPa)	$(3.0 ± 0.4) \times 10^{-1}$	4.3 ± 0.5	1.4 ± 0.3
Tensile strength (MPa)	0.7 ± 0.1	10.5 ± 1.1	3.4 ± 0.7
Tensile yield stress (MPa)	$(5.5 ± 1.4) \times 10^{-1}$	9.5 ± 1.0	2.3 ± 0.5
Maximum force ($\times 10^{-1}$ N)	1.3 ± 0.2	3.1 ± 0.3	2.1 ± 0.5

Consequently, the PLA35 film with wider fibers and lower porosity is the optimal film due to its desirable cooling performance and mechanical strength, as well as considerable wetting and adsorption properties to water, 1,2-propanediol and triglyceride, as summarized in Table 3.

Table 3. Summary of the physicochemical properties of the three electrospun/electrosprayed PEG/PLA films.

Physicochemical Property		PLA25	PLA35	PLA45
Morphology		fibers (0.14 µm)	fibers (0.80 µm)	particles
Thermal properties		good	good	good
Wetting properties (Contact Angle, °)	H_2O	88.4 ± 1.5	111.4 ± 0.8	100.9 ± 3.2
	1,2-propandiol	34.4 ± 1.7	36.7 ± 3.6	43.7 ± 2.8
	triglyceride	42.3 ± 1.2	51.2 ± 2.5	53.0 ± 3.5
Mechanical properties		worst	best	moderate

4. Conclusions

Herein, three PEG/PLA films with different morphologies were successfully prepared by varying the PLA content in the precursor solutions and by using the electrospinning or electrospray method. The morphology, phase composition, thermal properties, wetting properties and mechanical properties of the films were intensively studied, finding that the cooling performance of the PLA film was effectively improved by the addition of PEG due to its lower glass transition temperature and melting temperature. Moreover, the mechanical properties of the PEG/PLA films are greatly affected by their composition and morphology. It was demonstrated that the PLA35 film fabricated by the precursor electrospinning solution with a PLA content of 35 wt.% exhibited a desirable cooling performance, the best mechanical properties and acceptable wetting and adsorption properties towards water, 1,2-propanediol and triglyceride. As a result, the PLA35 film is a potential alternative film for HNB tobacco application.

Author Contributions: Conceptualization, W.K. and T.L.; methodology, X.L; software, X.L.; validation, M.M, B.L. and X.Z.; formal analysis, W.K.; investigation, W.K., X.L., M.M., B.L., X.Z. and T.L.; writing—original draft preparation, T.L.; writing—review and editing, W.K and T.L. All authors have read and agreed to the published version of the manuscript.

Funding: This research was funded by the Science and Technology Project of China Tobacco Hubei Industrial LLC, Grant No.: 2019420000340371.

Institutional Review Board Statement: Not applicable.

Informed Consent Statement: Not applicable.

Data Availability Statement: Data sharing not applicable.

Conflicts of Interest: The authors declare no conflict of interest.

References

1. Atlas, T.; Atlas, C.T. TitleCigarette Consumption, 2016. Available online: https://tobaccoatlas.org/topic/consumption/ (accessed on 10 July 2019).
2. Baran, W.; Madej-Knysak, D.; Sobczak, A.; Adamek, E. The influence of waste from electronic cigarettes, conventional cigarettes and heat-not-burn tobacco products on microorganisms. *J. Hazard. Mater.* **2020**, *385*, 121591. [CrossRef]
3. Benjamin, R.M. Exposure to tobacco smoke causes immediate damage: A report of the surgeon general. *Public Health Rep.* **2011**, *126*, 158–159. [CrossRef] [PubMed]
4. Centers for Disease Control and Prevention, National Biomonitoring Program, Tobacco. 2016. Available online: https://www.cdc.gov/biomonitoring/tobacco.html (accessed on 19 September 2019).
5. Shein, M.; Jeschke, G. Comparison of free radical levels in the aerosol from conventional cigarettes, electronic cigarettes, and heat-not-burn tobacco products. *Chem. Res. Toxicol.* **2019**, *32*, 1289–1298. [CrossRef]
6. Azzopardi, D.; Patel, K.; Jaunky, T.; Santopietro, S.; Camacho, O.M.; McAughey, J.; Gaça, M. Electronic cigarette aerosol induces significantly less cytotoxicity than tobacco smoke. *Toxicol. Mech. Method.* **2016**, *26*, 477–491. [CrossRef] [PubMed]

7. Simonavicius, E.; McNeill, A.; Shahab, L.; Brose, L.S. Heat-not-burn tobacco products: A systematic literature review. *Tob. Control.* **2019**, *28*, 582–594. [CrossRef] [PubMed]
8. Bentley, M.C.; Almstetter, M.; Arndt, D.; Knorr, A.; Martin, E.; Pospisil, P.; Maeder, S. Comprehensive chemical characterization of the aerosol generated by a heated tobacco product by untargeted screening. *Anal. Bioanal. Chem.* **2020**, *412*, 2675–2685. [CrossRef]
9. Kim, Y.-H.; An, Y.-J. Development of a standardized new cigarette smoke generating (SNCSG) system for the assessment of chemicals in the smoke of new cigarette types (heat-not-burn (HNB) tobacco and electronic cigarettes (E-Cigs)). *Environ. Res.* **2020**, *185*, 109413. [CrossRef] [PubMed]
10. Uchiyama, S.; Noguchi, M.; Takagi, N.; Hayashida, H.; Inaba, Y.; Ogura, H.; Kunugita, N. Simple determination of gaseous and particulate compounds generated from heated tobacco products. *Chem. Res. Toxicol.* **2018**, *31*, 585–593. [CrossRef]
11. Fu, Z.; Zhou, S.; Xia, L.; Mao, Y.; Zhu, L.; Cheng, Y.; Wang, A.; Zhang, C.; Xu, W. Juncus effusus fiber-based cellulose cigarette filter with 3D hierarchically porous structure for removal of PAHs from mainstream smoke. *Carbohyd. Polym.* **2020**, *241*, 116308. [CrossRef]
12. Branton, P.; Lu, A.-H.; Schüth, F. The effect of carbon pore structure on the adsorption of cigarette smoke vapour phase compounds. *Carbon* **2009**, *47*, 1005–1011. [CrossRef]
13. Li, G.; Yu, H.; Xu, L.; Ma, Q.; Chen, C.; Hao, Q.; Qian, Y. General synthesis of carbon nanocages and their adsorption of toxic compounds from cigarette smoke. *Nanoscale* **2011**, *3*, 3251–3257. [CrossRef]
14. Garlotta, D. A literature review of poly(lactic acid). *J. Polym. Environ.* **2001**, *9*, 63–84. [CrossRef]
15. Bao-shan, Y.; Tao, W.; Hao, W.; Jianbo, Z.; Geng, L.; Han, Z.; Jiao, X. Study on the influence of physical index of modified PLA filter rod on cigarette smoke index. In Proceedings of the 2020 3rd International Conference on Electron Device and Mechanical Engineering (ICEDME), Suzhou, China, 1–3 May 2020; pp. 494–496.
16. Ke, W.; Miao, M.; Liu, B.; Wu, Q.; Liu, T. Structural characteristics and properties of polylactic acid (PLA) and cellulose triacetate (CTA) fibers for heat-not-burn (HNB) cigarettes. *IOP Conf. Ser. Earth Environ. Sci.* **2021**, *719*, 042044. [CrossRef]
17. Tao, W.; Bao-shan, Y.; Hao, W.; Han, Z.; Jiao, X.; Jianbo, Z.; Geng, L. Safety Evaluation of Polylactic Aid Cgarette. In Proceedings of the 2020 3rd International Conference on Electron Device and Mechanical Engineering (ICEDME), Suzhou, China, 1–3 May 2020; pp. 222–224.
18. Zhang, M.; Li, X.H.; Gong, Y.D.; Zhao, N.M.; Zhang, X.F. Properties and biocompatibility of chitosan films modified by blending with PEG. *Biomaterials* **2002**, *23*, 2641–2648. [CrossRef]
19. Sundararajan, S.; Samui, A.B.; Kulkarni, P.S. Shape-stabilized poly(ethylene glycol) (PEG)-cellulose acetate blend preparation with superior PEG loading via microwave-assisted blending. *Sol. Energy* **2017**, *144*, 32–39. [CrossRef]
20. Şentürk, S.B.; Kahraman, D.; Alkan, C.; Gökçe, İ. Biodegradable PEG/cellulose, PEG/agarose and PEG/chitosan blends as shape stabilized phase change materials for latent heat energy storage. *Carbohyd. Polym.* **2011**, *84*, 141–144. [CrossRef]
21. Gök, Ö.; Alkan, C.; Konuklu, Y. Developing a poly(ethylene glycol)/cellulose phase change reactive composite for cooling application. *Sol. Energy Mater. Sol. Cell.* **2019**, *191*, 345–349. [CrossRef]
22. Peponi, L.; Sessini, V.; Arrieta, M.P.; Navarro-Baena, I.; Sonseca, A.; Dominici, F.; Gimenez, E.; Torre, L.; Tercjak, A.; López, D.; et al. Thermally activated shape memory effect on biodegradable nanocomposites based on PLA/PCL blend reinforced with hydroxyapatite. *Polym. Degrad. Stab.* **2018**, *151*, 36–51. [CrossRef]
23. Costa, L.M.M.; Mattoso, L.H.C.; Ferreira, M. Electrospinning of PCL/natural rubber blends. *J. Mater. Sci.* **2013**, *48*, 8501–8508. [CrossRef]
24. Liang, J.; Zhao, H.; Yue, L.; Fan, G.; Li, T.; Lu, S.; Chen, G.; Gao, S.; Asiri, A.M.; Sun, X. Recent advances in electrospun nanofibers for supercapacitors. *J. Mater. Chem. A* **2020**, *8*, 16747–16789. [CrossRef]
25. Parbey, J.; Xu, M.; Lei, J.; Espinoza-Andaluz, M.; Li, T.S.; Andersson, M. Electrospun fabrication of nanofibers as high-performance cathodes of solid oxide fuel cells. *Ceram. Int.* **2020**, *46*, 6969–6972. [CrossRef]
26. Bhardwaj, N.; Kundu, S.C. Electrospinning: A fascinating fiber fabrication technique. *Biotechnol. Adv.* **2010**, *28*, 325–347. [CrossRef] [PubMed]
27. Subbiah, T.; Bhat, G.S.; Tock, R.W.; Parameswaran, S.; Ramkumar, S.S. Electrospinning of nanofibers. *J. Appl. Polym. Sci.* **2005**, *96*, 557–569. [CrossRef]
28. Bognitzki, M.; Czado, W.; Frese, T.; Schaper, A.; Hellwig, M.; Steinhart, M.; Greiner, A.; Wendorff, J.H. Nanostructured fibers via electrospinning. *Adv. Mater.* **2001**, *13*, 70–72. [CrossRef]
29. Myndrul, V.; Vysloužilová, L.; Klápšťová, A.; Coy, E.; Jancelewicz, M.; Iatsunskyi, I. Formation and photoluminescence properties of ZnO nanoparticles on electrospun nanofibers produced by atomic layer deposition. *Coatings* **2020**, *10*, 1199. [CrossRef]
30. Ji, S.M.; Tiwari, A.P.; Kim, H.Y. Graphene oxide coated zinc oxide core–shell nanofibers for enhanced photocatalytic performance and durability. *Coatings* **2020**, *10*, 1183. [CrossRef]
31. Barhoum, A.; Pal, K.; Rahier, H.; Uludag, H.; Kim, I.S.; Bechelany, M. Nanofibers as new-generation materials: From spinning and nano-spinning fabrication techniques to emerging applications. *Appl. Mater. Today* **2019**, *17*, 1–35. [CrossRef]
32. Xue, J.; Wu, T.; Dai, Y.; Xia, Y. Electrospinning and electrospun nanofibers: Methods, materials, and applications. *Chem. Rev.* **2019**, *119*, 5298–5415. [CrossRef]
33. Aruna, S.T.; Balaji, L.S.; Senthil Kumar, S.; Shri Prakash, B. Electrospinning in solid oxide fuel cells—A review. *Renew. Sustain. Energy Rev.* **2017**, *67*, 673–682. [CrossRef]

34. Costa, L.M.M.; Bretas, R.E.S.; Gregorio, R., Jr. Effect of solution concentration on the electrospray/electrospinning transition and on the crystalline phase of PVDF. *Mater. Sci. Appl.* **2010**, *1*, 247–252. [CrossRef]
35. Bi, C.; Li, X.; Xin, Q.; Han, W.; Shi, C.; Guo, R.; Shi, W.; Qiao, R.; Wang, X.; Zhong, J. Effect of extraction methods on the preparation of electrospun/electrosprayed microstructures of tilapia skin collagen. *J. Biosci. Bioeng.* **2019**, *128*, 234–240. [CrossRef]
36. Haroosh, H.J.; Chaudhary, D.S.; Dong, Y. Electrospun PLA/PCL fibers with tubular nanoclay: Morphological and structural analysis. *J. Appl. Polym. Sci.* **2012**, *124*, 3930–3939. [CrossRef]
37. Casasola, R.; Thomas, N.L.; Trybala, A.; Georgiadou, S. Electrospun poly lactic acid (PLA) fibres: Effect of different solvent systems on fibre morphology and diameter. *Polymer* **2014**, *55*, 4728–4737. [CrossRef]
38. Altinisik, A.; Yurdakoc, K. Chitosan/poly(vinyl alcohol) hydrogels for amoxicillin release. *Polym. Bull.* **2014**, *71*, 759–774. [CrossRef]
39. Wang, Q.; Liu, P.; Liu, P.; Gong, T.; Li, S.; Duan, Y.; Zhang, Z. Preparation, blood coagulation and cell compatibility evaluation of chitosan-graft-polylactide copolymers. *Biomed. Mater.* **2014**, *9*, 015007. [CrossRef] [PubMed]
40. Henderson, M.A. The interaction of water with solid surfaces: Fundamental aspects revisited. *Surf. Sci. Rep.* **2002**, *46*, 1–308. [CrossRef]

MDPI
St. Alban-Anlage 66
4052 Basel
Switzerland
Tel. +41 61 683 77 34
Fax +41 61 302 89 18
www.mdpi.com

Coatings Editorial Office
E-mail: coatings@mdpi.com
www.mdpi.com/journal/coatings

www.ingramcontent.com/pod-product-compliance
Lightning Source LLC
LaVergne TN
LVHW070841170726
843515LV00005B/543